The Biology and Identification of the Coccidia (Apicomplexa) of Turtles of the World

The Biology and Identification of the Coccidia (Apicomplexa) of Turtles of the World

Donald W. Duszynski

Johnica J. Morrow

AMSTERDAM • BOSTON • HEIDELBERG • LONDON
NEW YORK • OXFORD • PARIS • SAN DIEGO
SAN FRANCISCO • SINGAPORE • SYDNEY • TOKYO

Academic Press is an imprint of Elsevier

Academic Press is an imprint of Elsevier
32 Jamestown Road, London NW1 7BY, UK
525 B Street, Suite 1800, San Diego, CA 92101-4495, USA
225 Wyman Street, Waltham, MA 02451, USA
The Boulevard, Langford Lane, Kidlington, Oxford OX5 1GB, UK

First published 2014

Notices
Knowledge and best practice in this field are constantly changing. As new research and experience
broaden our understanding, changes in research methods, professional practices, or medical
treatment may become necessary.

Practitioners and researchers must always rely on their own experience and knowledge in
evaluating and using any information, methods, compounds, or experiments described herein. In
using such information or methods they should be mindful of their own safety and the safety of
others, including parties for whom they have a professional responsibility.

To the fullest extent of the law, neither the Publisher nor the authors, contributors, or editors,
assume any liability for any injury and/or damage to persons or property as a matter of products
liability, negligence or otherwise, or from any use or operation of any methods, products,
instructions, or ideas contained in the material herein.

ISBN: 978-0-12-801367-0

Library of Congress Cataloging-in-Publication Data
A catalog record for this book is available from the Library of Congress

British Library Cataloguing-in-Publication Data
A catalogue record for this book is available from the British Library

For information on all Academic Press publications
visit our website at http://store.elsevier.com/

This book has been manufactured using Print On Demand technology. Each copy is produced
to order and is limited to black ink. The online version of this book will show color figures
where appropriate.

Working together
to grow libraries in
developing countries

ELSEVIER Book Aid International

www.elsevier.com • www.bookaid.org

This book is dedicated to Dr. William C. Marquardt (October 9, 1924—June 19, 2014), former Professor of Biology, Colorado State University, Fort Collins, CO. Dr. Marquardt was the mentor of DWD, which makes him the scientific grandfather of JJM. Most often during our journey through life, we don't take the time to tell people how much we value and love them, and later, the time never seems right or we don't take the time to make it right. Value and love are emotions based on interpersonal experience. This dedication will be a short journey through the experience of a graduate student (DWD) at Colorado State (1966–1970) under the direction of Bill Marquardt. It's time I took the time.

My first meeting with Dr. Marquardt was telling. He looked at my undergraduate transcript and asked me two questions: How did I ever get a degree in biology and how did I ever get accepted into graduate school? I left his office feeling pretty stupid (although these were legitimate questions, given my grades!). My fear of graduate school mounted when I entered "Ecology of Parasitism," my first graduate course that, of course, was taught by Bill (Fall Quarter, 1966). I still remember counting all those damn *Nematospiroides dubius* eggs, night after night, in a seemingly incessant stream of experiments he made us work on. In the Spring Quarter (1967) he had me TA the Histological Techniques (Z 146) class, a course I was taking and assisting simultaneously; that's one way to learn really quickly (blind leading the blind). But looking back, I wouldn't have wanted it any other way. Bill was an incredible source of information on every topic he talked about; I was amazed at his breadth and the depth of knowledge he possessed. Bill had a special gift of retaining the stuff he read and using it to problem solve. I thank him to this day for having been my teacher, for having frightened me into learning, and for lighting the fire of excitement about the coccidia that still burns within me today.

Besides being my teacher, Bill served as Chairman of my Committee on Studies for both the M.S. (1968) and Ph.D. (1970). He taught me *how* to write, the importance of writing well, and how to carry scholarship to its logical conclusion. He made me understand Koehler illumination, schooled me in how to use a photomicroscope, and taught me a skill very few possess: how to "see" the coccidia. He provided all of the fundamental groundwork for whatever claim I can make to being a scholarly person today. He let me work at my own speed and he trusted me. He also gave me strength when my personal life fell into crisis during my first year at CSU. I credit Bill, and the things he did for me during my graduate years, with primary responsibility for my being able to secure my first academic appointment, assistant professor of Biology, at the University of New Mexico (August, 1970).

As a faculty member at CSU, Dr. Marquardt was always one of the pillars of both the department and the university he served; a person of great knowledge and breadth, an individual of high ethical and moral standards, and a person with uncompromising integrity. He was a teacher, scholar, mentor, and friend, and he maintained his scholarly prowess until nearly the time of his death, with major reference/text books on biology of disease vectors, parasitology and vector biology, and other professional contributions.

Dr. Marquardt died while this book was *in press*. My biggest regret upon Bill's retirement, and now his passing, is that future generations of CSU students will not be able to experience what I got to experience: the Goodness of Bill Marquardt. I consider myself fortunate to have crossed his path when I did, and lucky to be able to call myself his friend and former student. Hopefully, I can instill in his academic granddaughter some of the same emotions.

TABLE OF CONTENTS

Preface and Acknowledgments xi

Chapter 1 Introduction 1
Turtles are Food, Pets, Lab Animals, and Majestic Creatures 1
Coccidia in Turtles: Perpetrators, Symptoms, and Disease..................3

Chapter 2 Suborder Cryptodira, Hidden-Necked Turtles 7
Family Chelydridae (Snapping Turtles) 7
Chelydra (3 *Eimeria*, 1 *Isospora* spp.) 7
Macrochelys (1 *Eimeria*) 14
Superefamily Testudinoidea..16
Family Emydidae (Pond, Box, Water Turtles)....................................16
Chrysemys (3 *Eimeria* spp.)...16
Emys (3 *Eimeria* spp.)...22
Glyptemys (2 *Eimeria* spp.) 27
Graptemys (3 *Eimeria* spp.) 31
Pseudemys (3 *Eimeria* spp.)...37
Terrapene (2 *Eimeria* spp.)..43
Trachemys (4 *Eimeria* spp.)..47
Family Testudinidae (Tortoises)...55
Chelonoidis (10 *Eimeria* spp.) 56
Gopherus (1 *Eimeria* sp.) 76
Testudo (1 *Eimeria*, 1 *Isospora* spp.) 79
Family Geoemydidae (Bataguridae) (Asian River, Leaf &
Roofed, Asian Box Turtles)..82
Batagur (1 *Eimeria* sp.)..82
Cuora (1 *Eimeria* sp.)..83
Cyclemys (2 *Eimeria* spp.)...85
Heosemys (1 *Eimeria* sp.)...88
Malayemys (1 *Eimeria* sp.)..90
Mauremys (1 *Eimeria* sp.)..92
Melanochelys (1 *Eimeria* sp.) 96
Pangshura (2 *Eimeria* spp.) 98

Superefamily Trionychoidea **102**
Family Trionychidae (Softshell Turtles) **103**
Apalone (7 *Eimeria* spp.) 103
Lissemys (4 *Eimeria* spp.) 117
Nilssonia (2 *Eimeria* spp.) 122
Superefamily Kinosternoidea **126**
Family Kinosternidae (Mud & Musk Turtles) **127**
Kinosternon (1 *Eimeria* sp.) 127
Superefamily Chelonioidea **129**
Family Cheloniidae (Sea Turtles) **129**
Caretta (1 *Eimeria* sp.) 129
Chelonia (1 *Caryospora* sp.) 131
Discussion and Summary **136**
Coccidia and Coccidiosis in Cryptodira 137
Endogenous Development 139
Treatment and Prevention 140
Host Specificity in Cryptodira Coccidia 140
Prevalence of Eimerians in Cryptodira 141
Archiving Biological Specimens for Future Study 142

Chapter 3 Suborder Pleurodira, Side-Necked Turtles **145**
Family Chelidae (Austro-American Sideneck Turtles) **145**
Mesoclemmys (1 *Eimeria* sp.) 146
Superefamily Pelomedusoidea **148**
Family Pelomedusidae (Afro-American Sideneck Turtles) **148**
Pelomedusa (1 *Eimeria* sp.) 148
**Family Podocnemididae (Madagascan Big-headed & American
Sideneck River Turtles)** **150**
Peltocephalus (1 *Eimeria* sp.) 150
Podocnemis (3 *Eimeria* spp.) 152
Discussion and Summary **158**

Chapter 4 *Cryptosporidium*, *Sarcocystis*, *Toxoplasma* in Turtles **161**
Cryptosporidium in Turtles 161
Sarcocystis in Turtles 163
Toxoplasma in Turtles 164

Chapter 5 *Species Inquirendae* in Turtles..**167**
Coccidium spp. (4) 167
Caryospora sp. (1)..169
Cryptosporidium spp. (9)..170
Eimeria spp. (5)..172
Mantonella sp. (1) 175
Sarcocystis spp. (8) 175

Chapter 6 Discussion and Summary **179**
Biodiversity...179
Variety of Oocyst Structures and Shape 180
Host Specificity...182
Pathology..183
Epidemiology..184
Treatment and Control...185
Archiving Biological Specimens...186
Closing Remarks 187

Tables **189**

Literature Cited..**203**

PREFACE AND ACKNOWLEDGMENTS

This book follows other similar reviews that have preceded it on the coccidian parasites of vertebrates (the majority of which still await discovery). For example, previous treatises have now been completed on the Amphibia (Duszynski et al., 2007), Serpentes (Duszynski & Upton, 2009), Chiroptera (bats) (Duszynski, 2002), Insectivora (moles, shrews) (Duszynski & Upton, 2000), Scandentia (tree shrews) (Duszynski et al., 1999), Primates (Prosimii, Anthropoidea) (Duszynski et al., 1999), and the Lagomorpha (rabbits, hares) (Duszynski & Couch, 2013). These studies illustrate that the vast majority of our knowledge about the coccidia of wild vertebrates has focused on mammals. We also know a great deal about the coccidia of some birds (e.g., domestic chickens) because of their importance as a food source for humans, but little is known about the coccidia and other parasites from reptiles. Here we contribute another piece of knowledge on the subject to help narrow that gap. We also note that the monoxenous coccidia make excellent organisms for the study of biodiversity because they are obtained from the feces of their hosts by noninvasive methods and their exogenous oocysts have relatively simple morphology, which can be studied easily with the light microscope.

We are grateful to Lee Couch, Department of Biology, The University of New Mexico (UNM), for her skillful assistance in scanning, "fixing," and archiving all of the line drawings and photomicrographs depicted in this book. This book would not be accurate in defining the host turtles without the seminal web archive, The Reptile Database, by P. Uetz and J. Hošek (eds.) (http://www.reptiledatabase.org/). We are grateful to many editors and/or publishers for giving us permission to utilize certain line drawings and photomicrographs from the original papers in which they appeared; these journals include: *Acta Protozoologica*, *Annals of Tropical Medicine and Hygiene* (now *Pathogens and Global Health*), *Canadian Journal of Zoology*, *Comparative Parasitology* (formerly *Journal of the Helminthological Society of Washington*), *Journal of Parasitology*, *Journal of Protozoology* (now *Journal of Eukaryotic Microbiology*), *Journal of Wildlife Diseases* (formerly *Bulletin of the Wildlife Disease Association*),

Memorias do Instituto Oswaldo Cruz, Parasite, Systematic Parasitology, Texas Journal of Science, and *Invertebrate Biology* (formerly *Transactions of the American Microscopical Society*). This book also would not have been possible without the dedicated work done by our numerous colleagues worldwide, who study the coccidia of turtles and publish their findings, and who gave us their permissions to use their line drawings and/or their photomicrographs; these individuals include, Drs./Professors D. Modrý (Czech Republic), P. Siroký (Czech Republic), C. McAllister (USA), R. Lainson (Brazil), and L. Couch (USA).

Finally, this book would not have been produced and available to the parasitological community without the dedicated, professional staff at Elsevier, especially, Linda Versteeg-buschman, Acquisitions Editor; Halima Williams, Editorial Project Manager, Life Sciences; and Anusha Sambamoorthy, Project Manager, Book Production, Chennai.

Donald W. Duszynski
Professor Emeritus of Biology,
The University of New Mexico,
Albuquerque, NM

Johnica J. Morrow
School of Natural Resources,
University of Nebraska-Lincoln,
Lincoln, NE

For information, questions, comments, or corrections contact either D.W. Duszynski at eimeria@unm.edu or Johnica J. Morrow at johnica@huskers.unl.edu.

CHAPTER 1

Introduction

This treatise on coccidia species known from turtles has several prede-
cessors including, *The Coccidia of Snakes of the World* (Duszynski &
Upton, 2009) and *The Biology and Identification of the Coccidia of
Rabbits of the World* (Duszynski & Couch, 2013). Like the others,
it is intended to be the most comprehensive discourse, to date, describ-
ing the structural and biological knowledge of all coccidian parasites
that infect turtles. These protists (Phylum Apicomplexa) seem relatively common in turtles and are represented by about 71 species
that fit taxonomically into five genera in three families that include
Cryptosporidiidae Léger, 1911 (*Cryptosporidium*), Eimeriidae Minchin,
1903 (*Caryospora, Eimeria, Isospora*), and Sarcocystidae Poche, 1913
(*Sarcocystis*). An overview of the general biology, taxonomy, life
cycles, and numbers of species of eimeriid and cryptosporid coccidia
from wild mammals was published a decade ago (Duszynski & Upton,
2001), and monographic works on the coccidia of certain selected vertebrate groups also are available including: Amphibia (Duszynski
et al., 2007); Chiroptera (Duszynski, 2002); Insectivora (Duszynski &
Upton, 2000); Marmotine squirrels (Rodentia) (Wilber et al., 1998);
and Primates and Scandentia (Duszynski et al., 1999). No such review
exists for the coccidia of turtles. Here we strive to resolve that void
because turtles have a long and important history shared with humans
and many (>50%) of their extant species are threatened with extinction as human populations continue to increase on Earth.

TURTLES ARE FOOD, PETS, LAB ANIMALS, AND MAJESTIC CREATURES

Almost everyone can recognize turtles because of their shells. These
are remarkable and distinguishing structures that enclose the body of
the entire animal in a bony case that only opens at the front and the
rear. The shell, of course, helps to protect them from natural enemies,
but it also has limited their morphological diversity; there are terrestrial and aquatic turtles, but no turtles can climb (arboreal) or fly.

The Biology and Identification of the Coccidia (Apicomplexa) of Turtles of the World.
DOI: http://dx.doi.org/10.1016/B978-0-12-801367-0.00001-0

The habits of turtles often may be deduced by the appearance of their shells. Terrestrial turtles (e.g., tortoises and box turtles) generally have high, domed shells and stout limbs, whereas aquatic turtles usually have relatively flat shells (for less resistance in water) and webbed feet. However, some aquatic forms (e.g., mud and musk turtles) spend more time walking on the bottom than they do swimming, and these may have more distinct dome-shaped shells than those that swim quickly to capture prey and avoid predation. Other aquatic turtles have even more specialized soft shells that lack a bony layer, while still others have greatly reduced dermal bones and the stiff dermal scales, which have been replaced by a flexible covering of skin (Pough et al., 2004).

Sometimes the term Testudines is used by biologists to encompass all turtles, including both their extinct ancestors and the two major groups of living (extant) turtles (Cryptodira and Pleurodira) and their descendants. The term Chelonia is widely used for the extant clades only. All extant turtles belong to one of these two groups (clades) that are taxonomically referred to as suborders. The Cryptodira and the Pleurodira are distinguished from each other by the way they retract their necks. Turtles placed into the Cryptodira retract their neck in a vertical plane, whereas members of the Pleurodira (side-necked turtles) retract their necks in a horizontal or sideway plane. Nonetheless, because of the many unique structural features (shell, skeleton, skull bones, others), the monophyly of turtles has never been in serious question by herpetologists (Pough et al., 2004).

Turtles and humans share a long association, with the former being used as gourmet food, as subjects in both oriental and traditional medicine and contemporary medical research, their fat as a base for cosmetics, their shells as jewelry and, of course, millions of humans have them as pets. There are currently 328 distinct turtle species recognized by herpetologists. However, the number of turtle taxa becomes more complex because 56 of those species are polytypic (have more than one immediately subordinate taxon), which introduces 124 additional recognized subspecies (Rhodin et al., 2010). Thus, herpetologists now recognize 452 taxa of modern tortoises and turtles, of which 10 taxa (8 species, 2 subspecies) are now considered extinct. In their 2010 update on turtles of the world, Rhodin et al. (2010) listed 156 of the 328 species (48%) as Threatened, with 90 (27%) as Critically Endangered or Endangered on the International Union for the Conservations of Nature (IUCN) Red List. When they included Extinct-in-the-Wild and Extinct species, their data suggested that at least 50% of all modern turtle and tortoise species either are already extinct or threatened with extinction. In other words, turtles are threatened with extinction at a much higher risk than almost all other vertebrate species. The threats that turtles face include, but are not limited to, long-term unsustainable exploitation, habitat destruction, overharvesting for consumption, and the international pet trade.

COCCIDIA IN TURTLES: PERPETRATORS, SYMPTOMS, AND DISEASE

This review primarily is concerned with the eimeriid coccidia found in turtles, but mention also will be made of genera and species in two other protist families, Sarcocystidae (*Sarcocystis, Toxoplasma* spp.) and Cryptosporididae (*Cryptosporidium* sp.). All species in these genera are single-celled, intracellular, eukaryotic parasites (Protozoa: Apicomplexa) that pass a highly resistant propagule, the oocyst, in the feces of their definitive host. Most of these parasites develop in the epithelial cells of the gastrointestinal tract, but a few develop in other tissues and organs. Their life cycles are complex and include both asexual reproduction (merogony), which can produce extremely high numbers of individuals, and sexual reproduction (gamogony) that allows genetic recombination. After fertilization takes place and oocyst wall formation and development is completed, the oocyst leaves the host epithelial cell, destroying it, and usually (but not always) needs a period of time outside the host to undergo its final developmental process, sporogony. After sporogony, the oocyst becomes infective to the next host that may encounter this stage and ingest it.

The majority of the species summarized in our review are in the Eimeriidae (*Eimeria, Isospora, Caryospora* spp.). As best we know, all of these parasites are homoxenous in their development; that is, they have a single host, direct (host-to-host) life cycle. *Cryptosporidium* species also have direct life cycles, but *Sarcocystis* species are heteroxenous in that they require two hosts, a definitive host that discharges the oocysts, and an intermediate host that contains infective tissue stages/cysts that must be eaten by the definitive host to complete the life cycle. *Toxoplasma* species can be either homoxenous (only in felids) or heteroxenous in their development. Even though their intracellular development kills their hosts' epithelial or endothelial cells, most coccidia are considered nonpathogenic; however, as we will learn in the following chapters, we know so little about the complete development and life cycles of the majority of known species, that we really don't know what levels of pathogenicity may exist in turtle coccidia. Still, there are exceptions to these gaps in our understanding, as noted in later chapters.

The most distinguishing feature of the coccidian life cycle that allows diagnosis, for example, by a veterinarian, is the structure of the sporulated oocyst; oocysts of most coccidians (e.g., *Caryospora, Eimeria, Isospora, Toxoplasma* spp.) leave their host unsporulated and need molecular oxygen and usually a temperature different than the body temperature of their host to undergo sporulation. A few, however, like the *Cryptosporidium* and *Sarcocystis* species, have completely sporulated oocysts and/or sporocysts that leave their host in its feces. In the most general terms, the oocysts of the genera covered in this review are distinguished as follows: *Caryospora* has one sporocyst with eight sporozoites, and it leaves the host unsporulated; *Cryptosporidium* does not have a sporocyst, but has four sporozoites in the oocyst, and it leaves the host completely sporulated; *Eimeria* has four sporocysts, each with two sporozoites, and it leaves the host unsporulated; *Isospora* has two sporocysts, each with four sporozoites, and it leaves the host unsporulated; *Sarcocystis* has very thin-walled oocysts that completely sporulate, but which usually break within the gut releasing their two sporocysts, each with four sporozoites; *Toxoplasma* has very small (≤ 10 μm wide) oocysts, with two sporocysts, each with four sporozoites, and it leaves the definitive host (only felids) unsporulated.

When comparing coccidian oocysts that may at first glance seem quite similar, it soon becomes clear that some structural characteristics, both qualitative and quantitative, can be very useful for species identification. We have found that the most important oocyst characters and their standardized abbreviations for oocyst–sporocyst structures are those proposed by Wilber et al. (1998), which have now become the standard for descriptive parameters of oocysts: *Oocyst Characters*: length (*L*), width (*W*), their ranges and ratio (*L/W*), micropyle (M), residuum (OR), polar granules (PG); *Sporocyst Characters*: length (*L*), width (*W*), their ranges and ratio (*L/W*), Stieda body (SB), substieda body (SSB), parastieda body (PSB), residuum (SR); *sporozoite* (SZ) *characters*: refractile bodies (RB) and nucleus (N). All measurements noted in this book are given in micrometers (μm) with size ranges in parentheses following the means, when available.

When a sporulated (infective) oocyst is ingested by the appropriate host, the sporozoites are released in the host gastrointestinal tract and seem to have a predilection to enter certain epithelial cells only in certain specific parts of the gut. Once inside the cell, they round up and develop into a trophozoite that feeds and grows; the nucleus fragments and cytokinesis takes place, a process called merogony (asexual multiple fission), that produces tens to hundreds to thousands of genetically identical merozoites, each capable of infecting other epithelial cells, destroying the initial host cell in the process. Merogony seems to be genetically programmed, occurring a specific number of times, depending on the species. Eventually, the final generation of merozoites enters host cells and produce either microgametocytes that release microgametes or macrogametocytes, that can go on to become viable oocysts once fertilized. After fertilization and formation of the oocyst wall, the oocysts leave the host. Some are unsporulated (e.g., *Eimeria* spp.) and need time as well as the appropriate conditions outside of the host to undergo the sporulation process before they become infective to another suitable host. Others (e.g., *Sarcocystis* spp.) sporulate within the gut cell of the host in which gamogony occurs and discharge oocysts/sporocysts that are immediately infective to the next available host. In all cases, it is important to remember that coccidia infections are transient, unlike helminth infections that may be retained by their hosts for years or even decades. That is, coccidia infections produce relatively brief infections consisting of a finite number of endogenous cycles of multiplication, usually in the intestinal tract or associated ducts and/or organs; the cycle is then terminated by the formation of oocysts, which are passed from the host in their feces, thus ending the infection.

Caryospora, Eimeria, and *Isospora* species are thought to be the most host specific of the coccidia, being able to infect host species within a single host genus or within closely related (sister) genera. *Toxoplasma* species seem to be almost ubiquitous for intermediate vertebrate hosts in which it can form tissue cysts via various asexual processes, while gamogony can only occur in felids in these species.

Sarcocystis species require two hosts, a carnivore and an omnivore, with the carnivore being the only host that can discharge oocysts because it is the only host in which gamogony will occur. For more details on coccidian life cycles and structures in reptile coccidians, see reviews by Duszynski and Upton (2001, 2009).

In our review we have used the most current classification schemes for turtles found on the World Wide Web (Anonymous, 2010; Rhodin et al., 2010; Uetz & Hošek, 2013). The families and genera of turtles listed in this book are in alphabetical order and no attempt is made to show phylogenetic relationships of the host species from which coccidia have been described. The order Testudines (turtles) consists of two extant suborders, 14 families with 92 genera, and these comprise 328 species at present. Coccidia have been described from 10 of 14 (71%) families, 32 of 92 (35%) genera, and only 64 of 328 (19.5%) turtle species. Based on our evaluation of every species in the published literature we document, as valid named species, 66 *Eimeria*, three *Isospora*, one *Caryospora*, and one *Sarcocystis* spp. In addition, there are an additional nine *Cryptosporidium*, five *Eimeria*, eight *Sarcocystis*, one *Caryospora*, one *Mantonella*, and four "*Coccidium*" species reported from turtles, but their descriptions are so inadequate that these must be considered as *Species Inquirendae* until further study and data can be had to determine their validity and true identity.

CHAPTER 2

Suborder Cryptodira, Hidden-Necked Turtles

This suborder includes most living tortoises and turtles. The main unifying character of this group is that they lower their necks and pull their heads straight back into their shells. Their neck bones are wide and flat with a number of the joints able to flex through some 170° to allow the neck to fold back onto itself. This structure allows for little sideways movement.

FAMILY CHELYDRIDAE, SNAPPING TURTLES, 2 GENERA, 4 SPECIES

Genus *Chelydra* Schweigger, 1812 (3 Species)

Eimeria chelydrae Ernst, Stewart, Sampson, & Fincher, 1969

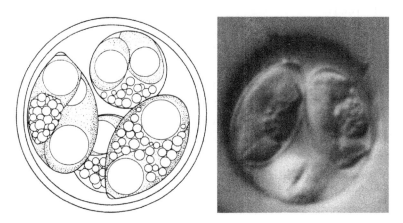

Figures 2.1, 2.2 Line drawing of the sporulated oocyst of Eimeria chelydrae *from Ernst et al. (1969), with permission from the* Journal of Wildlife Diseases. *Photomicrograph of a sporulated oocyst of* E. chelydrae *from McAllister et al. (1990b), with permission from the* Canadian Journal of Zoology *and from the senior author.*

Type host: *Chelydra serpentina* (L., 1758), Common Snapping Turtle.
Type locality: NORTH AMERICA: USA: Georgia, farm pond near Tifton.

The Biology and Identification of the Coccidia (Apicomplexa) of Turtles of the World.
DOI: http://dx.doi.org/10.1016/B978-0-12-801367-0.00002-2

Other hosts: None to date.

Geographic distribution: NORTH AMERICA: USA: Arkansas, Georgia.

Description of sporulated oocyst: Oocyst shape: spheroidal or sub-spheroidal, rarely ellipsoidal; number of walls: 1; wall characteristics: smooth, colorless to light blue, ~0.6–0.8 wide; L × W (N = 50): 15.2 × 14.4 (13–17 × 12–17); L/W ratio: 1.0 (1.0–1.2); M, OR, PG: all absent. Distinctive features of oocyst: Spheroidal shape, small size, and lack of M, OR, and PG.

Description of sporocyst and sporozoites: Sporocyst shape: ovoidal; L × W (N = 50): 9.6 × 5.6 (8–10 × 5–6); L/W ratio: (1.7); SB: present, small, at pointed end of sporocyst; SSB, PSB: both absent; SR: present; SR characteristics: granules of various sizes in a mass surrounded by a thin membrane or scattered throughout sporocyst; SZ: fusiform, lying lengthwise in the sporocysts, usually with RB in each end. Distinctive features of sporocyst: sometimes with a membrane-bound SR and SZ with two RB each.

Prevalence: In 1/1 (100%) of the type host; in 3/6 (50%) of the same host species from Arkansas, and 1/3 (33%) from Texas (McAllister et al., 1994).

Sporulation: Exogenous. Turtle feces with oocysts were collected in Georgia, placed in 2.5% $K_2Cr_2O_7$ solution, and mailed to the USDA Parasitology Research Lab, Auburn, Alabama, where partially sporulated and a few sporulated oocysts were seen. Oocysts were then allowed to continue to sporulate for 1 week at 20–24°C and later stored at 4°C for 3 weeks before being studied.

Prepatent and patent periods: Unknown oocysts were collected from the feces.

Site of infection: Unknown.

Endogenous stages: Unknown.

Cross-transmission: None to date.

Pathology: Unknown.

Materials deposited: None.

Entymology: The specific epithet reflects the genus name of the turtle type host.

Remarks: This was the first eimerian described from snapping turtles and it is clearly different from the other two eimerians from this host species in size and in structure of its oocysts and sporocysts.

Eimeria filamentifera **Wacha & Christiansen, 1979a**

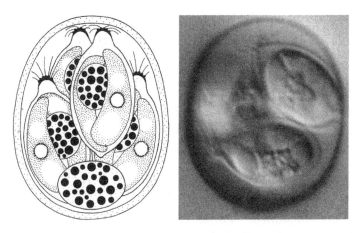

Figures 2.3, 2.4 Line drawing of the sporulated oocyst of Eimeria filamentifera *from Wacha and Christiansen (1979a), with permission from John Wiley & Sons, Ltd. holder of the copyright for the* Journal of Eukaryotic Microbiology *(formerly* Journal of Protozoology*). Photomicrograph of a sporulated oocyst of* E. filamentifera *from McAllister et al. (1990b), with permission from the* Canadian Journal of Zoology *and from the senior author.*

Type host: *Chelydra serpentina* (L., 1758), Common Snapping Turtle.
Type locality: NORTH AMERICA: USA: Iowa, Lee County, Shimek State Forest.
Other hosts: None to date.
Geographic distribution: NORTH AMERICA: USA: Arkansas, Iowa.
Description of sporulated oocyst: Oocyst shape: slightly ovoidal to ellipsoidal; number of walls: 1; wall characteristics: smooth outer surface, ~1.0 thick; L × W (N = 25): 23.2 × 18.6 (19−27 × 14.5−21); L/W ratio: 1.2 (1.1−1.6); M, PG: both absent; OR: ellipsoidal or spheroidal, membrane-bound, 7.4 × 6.1 (4−10 × 4−9) with numerous spheroidal granules, ~1.0 (0.5−1.5) each. Distinctive features of oocyst: Slightly ovoidal shape with a thin, smooth wall.
Description of sporocyst and sporozoites: Sporocyst shape: ellipsoidal; L × W: 14.0 × 7.7 (11.5−16 × 6.5−9.5); L/W ratio: (1.8); SB: present on a small, knob-like or neck-like protrusion of sporocyst wall, and with a tuft of filaments, each arising from a separate papilla; SSB: absent; PSB: absent, but there is a knob-like protrusion of the sporocyst wall opposite the SB; SR: present; SR characteristics: spheroidal or ellipsoidal membrane-bounded body, 5.1 × 4.7 (3−9 × 2.5−7.5), containing numerous spheroidal granules, ~1.0 (0.5−1.5) each; SZ: with an ellipsoidal, posterior RB, ~3.7 × 2.6, and a smaller, anterior spheroidal RB, ~2.0; N, ~2.0

wide, occasionally visible in middle of SZ; SZ: ~12.0 × 4.0, *in situ*. Distinctive features of sporocyst: neck-like end with an SB that possesses filaments each of which arises from a separate papilla.

Prevalence: In 2/2 (100%) of the type host; in 1/6 (17%) of the same host from Arkansas (McAllister et al., 1994).

Sporulation: Exogenous. Oocysts sporulated in 1 week when placed in 2.5% aqueous (w/v) $K_2Cr_2O_7$ solution at ~22°C and then stored for 90 days in a refrigerator at ~5°C.

Prepatent and patent periods: Unknown.

Site of infection: Unknown, oocysts recovered from feces.

Endogenous stages: Unknown.

Cross-transmission: None to date.

Pathology: Unknown.

Materials deposited: Drake University, Des Moines, Iowa (Collection numbers TWN3 and DPJ9).

Entymology: The specific epithet refers to the filamentiferous nature of the SB.

Remarks: *E. filamentifera* differs from the other two eimerians described from snapping turtles by the unique shape of its sporocysts, each having papillose filament-bearing SB and a small knob-like projection of the sporocyst wall at the end opposite the SB.

Eimeria serpentina McAllister, Upton, & Trauth, 1990b

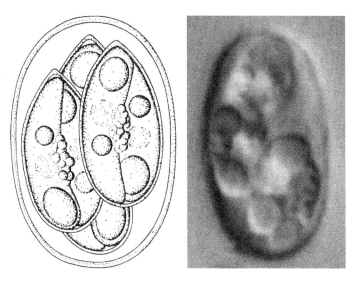

Figures 2.5, 2.6 Line drawing of the sporulated oocyst of Eimeria serpentina *from McAllister et al. (1990b), with permission from the* Canadian Journal of Zoology *and from the senior author. Photomicrograph of a sporulated oocyst of* E. serpentina *from McAllister et al. (1990b), with permission from the* Canadian Journal of Zoology *and from the senior author.*

Type host: *Chelydra serpentina serpentina* (L., 1758), Common Snapping Turtle.

Type locality: NORTH AMERICA: USA: Arkansas, Boon County, 24.5 km W Yellville, off route 412 (T18N, R18N, Sec. 9).

Other hosts: None to date.

Geographic distribution: NORTH AMERICA: USA: Arkansas, Texas.

Description of sporulated oocyst: Oocyst shape: ellipsoidal; number of walls: 1; wall characteristics: smooth, thin, single-layered, ~0.3 thick, that wrinkles easily in sucrose solution; L × W (N = 30): 12.8 × 8.1 (11−15 × 7−10); L/W ratio: 1.6 (1.4−1.9); M, OR, PG: all absent. Distinctive features of oocyst: thin wall that wrinkles easily in sucrose solution.

Description of sporocyst and sporozoites: Sporocyst shape: ellipsoidal; L × W (N = 20): 7.6 × 4.1 (6−9 × 4−5); L/W ratio: 1.8 (1.5−2.1); SB: present, small, at tapered end of sporocyst; SSB, PSB: both absent; SR: present; SR characteristics: rarely compact, more often scattered among SZ; SZ: sausage-shaped, 7.3 × 2.2 (6−8 × 2−3) *in situ*, arranged head-to-tail within the sporocyst; each SZ with a spheroidal anterior RB, 1.3 (1−2) and a subspheroidal−ellipsoidal posterior RB, 1.9 × 1.7 (1−2 × 1−2), with an N between them. Distinctive features of sporocyst: None.

Prevalence: In 4/9 (44%) of the type host, including 2/6 (33%) from Arkansas and 2/3 (67%) from Texas.

Sporulation: Exogenous. Oocysts were passed unsporulated and became fully sporulated within 4−6 days at 22°C in petri dishes containing a thin layer of tap water supplemented with antibiotics.

Prepatent and patent periods: Unknown.

Site of infection: Unknown. Oocysts found in feces and intestinal contents.

Endogenous stages: Unknown.

Cross-transmission: None to date.

Pathology: Unknown.

Materials deposited: Oocysts in 10% formalin in the USNPC, No. 80759. Symbiotype host in the Arkansas State University Museum of Zoology, ASUMZ 13108.

Entymology: The specific epithet of this eimerian is derived from the specific epithet of the turtle type host.

Remarks: Oocysts of this species clearly differ from the other two eimerians described from snapping turtles in size and shape of both their oocysts and sporocysts. This species may represent the unnamed eimerian of Wacha and Christiansen (1980) from the same host species in Iowa. They reported an *Eimeria* species with ovoidal–ellipsoidal oocysts, 15×9, and ovoidal sporocysts, 8.5×4.0. Like *E. serpentina*, their eimerian had an SR, but not an OR. Thus, McAllister et al. (1990) argued that the *Eimeria* species of Wacha and Christiansen (1980) should be provisionally regarded as a synonym of *E. serpentina*.

Isospora chelydrae McAllister, Upton, & Trauth, 1990b

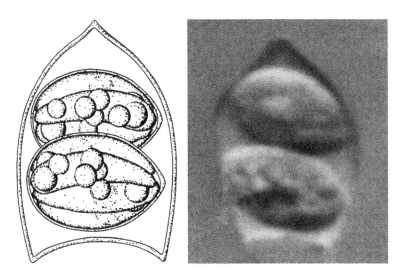

Figures 2.7, 2.8 Line drawing of the sporulated oocyst of Isospora chelydrae *from McAllister et al. (1990b), with permission from the* Canadian Journal of Zoology *and from the senior author. Photomicrograph of a sporulated oocyst of* I. chelydrae *from McAllister et al. (1990b), with permission from the* Canadian Journal of Zoology *and from the senior author.*

Type host: *Chelydra serpentina serpentina* (L., 1758), Common Snapping Turtle.
Type locality: NORTH AMERICA: USA: Arkansas, Carroll County, 11.3 km E Marble, off route 412 (T17N, R23W, Sec. 6).
Other hosts: None to date.
Geographic distribution: NORTH AMERICA: USA: Arkansas.
Description of sporulated oocyst: Oocyst shape: asymmetrical; number of walls: 1; wall characteristics: smooth, thin, single-layered, $\sim 0.2 - 0.4$ thick; $L \times W$ ($N = 25$): 9.6×6.8 ($9 - 11 \times 6 - 8$); L/W ratio: 1.4 (1.2–1.6); M, OR, PG: all absent. Distinctive features of oocyst: has one conical projection on one side of the oocyst and

two conical projections on the opposite side of the oocyst, each 1.5–1.6 long.

Description of sporocyst and sporozoites: Sporocyst shape: ellipsoidal; L × W (N = 25): 5.8 × 4.0 (5–7 × 4); L/W ratio: 1.5 (1.3–1.7); SB: present, small, at pointed end of sporocyst; SSB, PSB: both absent; SR: present; SR characteristics: spheroidal, ovoidal, or scattered; when intact 2.9 × 2.5 (2–4 × 2–3), composed of a cluster of two to eight large globules, each 0.8–1.4 wide, not membrane bound; SZ: 5.6 × 1.1 (5–6 × 1–2) *in situ*, arranged head-to-tail within the sporocyst. SZ with a spheroidal posterior RB and a central N. Distinctive features of sporocyst: both the RB and N are very faint and difficult to visualize.

Prevalence: In 3/9 (33%) of the type host in the original description; in 3/6 (50%) of the same host (McAllister et al., 1994) also in Arkansas.

Sporulation: Endogenous. Oocysts were passed fully sporulated.

Prepatent and patent periods: Unknown.

Site of infection: Unknown. Oocysts found in feces and intestinal contents.

Endogenous stages: Unknown.

Cross-transmission: None to date.

Pathology: Unknown.

Materials deposited: Oocysts in 10% formalin in the USNPC, No. 80760. Symbiotype host in the Arkansas State University Museum of Zoology, ASUMZ 13107.

Entymology: The specific epithet of this eimerian is derived from the genus name of the turtle type host.

Remarks: There are only two other recorded isosporans in turtles; one is *Isospora testudae* described from Horsfield's tortoise, *Testudo horsfieldi*, in Ubekistan (Davronov, 1985) and the other is *Isospora rodriguesae* reported from the yellow-footed tortoise, *Chelonoidis denticulata*, in Brazil (Lainson et al., 2008); both genera are in the Testudinidae. Davronov (1985) found 17/63 (27%) *T. horsfieldi* from Uzbekistan (former USSR) to be shedding oocysts of *I. testudae*. The oocysts of that isosporan are spheroidal and large, 25.6 (22–29), with ovoidal sporocysts 15–19 × 10–15 (Davronov, 1985), whereas those of *I. chelydrae* are asymmetrical with a thin wall (<1.0) and are much smaller, 9.6 × 6.8 (9–11 × 6–8) containing sporocysts that are also very small, 5.8 × 4.0 (5–7 × 4). Lainson et al. (2008) found oocysts of *I. rodriguesae* in only 1/8 (12.5%)

yellow-footed tortoises; these were broadly ellipsoidal to subspheroidal and had a smooth, thick wall (~1.0) and oocysts that measured 24.5 × 21.0 (23−26 × 22), lacked both M and OR, but had a distinct PG present; all characteristics clearly distinguish these three isosporan species from turtles.

Genus *Macrochelys* Gray, 1856 (Monospecific)
Eimeria harlani Upton, McAllister, & Trauth, 1992

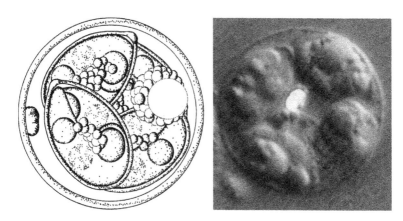

Figures 2.9, 2.10 Line drawing of the sporulated oocyst of Eimeria harlani *from Upton et al. (1992), with permission from* Comparative Parasitology *(formerly* Proceedings of the Helminthological Society of Washington*) and from C.T. McAllister. Photomicrograph of a sporulated oocyst of* E. harlani *from Upton et al. (1992), with permission from* Comparative Parasitology *(formerly* Proceedings of the Helminthological Society of Washington*) and from C.T. McAllister.*

Type host: *Macrochelys temminckii* Troost, 1835 (subadult) Alligator Snapping Turtle.
Type locality: NORTH AMERICA: USA: Arkansas, Jackson County, Black River, 9.7 km west of Swifton.
Other hosts: None to date.
Geographic distribution: NORTH AMERICA: USA: Arkansas.
Description of sporulated oocyst: Oocyst shape: spheroidal to subspheroidal; number of walls: 1; wall characteristics: single, thin layer ~0.5 thick; L × W: 13.0 × 12.6 (10−14 × 10−14); L/W ratio: 1.0 (1.0−1.1); M: absent; OR: present as delicate spheroidal to subspheroidal mass of granules usually surrounding a vacuole-like structure; PG: present. Distinctive features of oocyst: round with a distinctive OR and a distinct PG.

Description of sporocyst and sporozoites: Sporocyst shape: ovoidal with a thin wall ~0.4 thick; L × W: 8.9 × 5.2 (8−10 × 5−6); L/W ratio: 1.7 (1.6−1.8); SB: present as thickening at one end of sporocyst; SSB, PSB: both absent; SR: present; SR characteristics: either as 3−12 scattered granules or as compact mass 3.1 × 2.9 (2−4 × 2−3); SZ (N = 20): elongated sausage-shaped, 10.5 × 2.3 (8−12 × 2−3) *in situ*, arranged head-to-tail within sporocyst; posterior ends of SZ reflected back along one end of sporocyst and each SZ has a spheroidal anterior RB (N = 20), 1.8 (1−2) wide and a spherical posterior RB (N = 20), 2.2 (2−3) wide; N of SZ located between RBs. Distinctive features of sporocyst: perfect ovoidal shape with an SB, SR, and SZs with two RBs.

Prevalence: In 1/1 (100%) of the type host; in 1/4 (25%) of the same host species, also in Arkansas (McAllister et al., 1994).

Sporulation: Exogenous. All oocysts were passed unsporulated and became fully sporulated within 1 week at ~23°C.

Prepatent and patent periods: Unknown.

Site of infection: Unknown. Oocysts recovered from intestinal contents and feces.

Endogenous stages: Unknown.

Cross-transmission: None to date.

Pathology: None observed in the two infected hosts examined.

Materials deposited: Host voucher specimen: Arkansas State University Museum of Zoology as ASUMZ 17616; Syntypes of sporulated oocysts in 10% formalin: US National Museum, Beltsville, Maryland 20705 as USNM No. 82005.

Entymology: The specific epithet is given in honor of Richard Harlan (1796−1843), American vertebrate paleontologist and comparative anatomist, who described the alligator snapping turtle in 1835, originally under the name *Chelonura temminckii* (see Bour, 1987).

Remarks: No species of *Eimeria* were reported from alligator snapping turtles prior to this description, nor have any been reported since this description appeared. Sporulated oocysts of *E. harlani* also are unlike any reported to date from the family Chelydridae. The only sporulated oocysts that are somewhat similar are those of *E. serpentina*, but they have smaller oocysts and lack an OR.

SUPERFAMILY TESTUDINOIDEA

FAMILY EMYDIDAE, POND, BOX, WATER TURTLES, 11 GENERA, 50 SPECIES

Genus *Chrysemys* Gray, 1844 (Monospecific)

Eimeria chrysemydis Deeds & Jahn, 1939

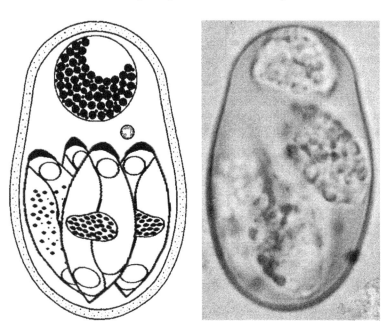

Figures 2.11, 2.12 Line drawing of the sporulated oocyst of Eimeria chrysemydis *(original) modified from Deeds and Jahn (1939), and from Wacha and Christiansen (1976). Photomicrograph of a sporulated oocyst of* E. chrysemydis *from Wacha and Christiansen (1976) with permission from John Wiley & Sons, Ltd., holder of the copyright for the* Journal of Eukaryotic Microbiology *(formerly* Journal of Protozoology*).*

Type host: *Chrysemys picta belli* (Gray, 1831) (syn. *Chrysemys bellii marginata* Conant, 1938), Western Painted Turtle.

Type locality: NORTH AMERICA: USA: Iowa, Lake West Okoboji.

Other hosts: *Graptemys caglei* Haynes & McKnown, 1974, Cagle's Map Turtle; *Graptemys geographica* (La Sueur, 1817), Common Map Turtle; *Trachemys gaigeae* (Hartweg, 1939), Big Bend Slider; *Trachemys scripta elegans* (Wied, 1838), Red Eared Slider.

Geographic distribution: NORTH AMERICA: USA: Arkansas, Iowa, New Mexico, Texas.

Description of sporulated oocyst: Oocyst shape: ovoidal to slightly pear-shaped; number of walls: 1 or 2, ∼1 thick; wall characteristics: smooth, slightly yellow, transparent membrane(s) with a thin fracture line around circumference of oocyst, ∼20% of total length below slightly pointed end; L × W: 23 × 15 (21–27 × 13–18) and L/W ratio: 1.5 (Deeds & Jahn, 1939) or L × W: 27.6 × 17.0 (24–30 × 14.5–18.5) and L/W ratio: 1.6 (1.3–1.9) (Wacha & Christiansen, 1976); M: absent; PG: present or absent; OR: present; OR characteristics: ellipsoidal to spheroidal, membrane-bound body, 5–9, with a distinct vaculated area surrounded by numerous spheroidal granules, each ∼0.5 and always near pointed end of oocyst. Distinctive features of oocyst: pear-shape, large OR near polar end, and thin fracture line around circumference of oocyst near pointed end.

Description of sporocyst and sporozoites: Sporocyst shape: spindle-shaped or narrowly ellipsoidal and clustered at rounded end opposite OR; L × W: 12–14 × 5–8 (Deeds & Jahn, 1939) or 13.6 × 6.5 (12–14.5 × 5–8); L/W ratio: 2.1; SB: present, thin, convex; SSB, PSB: both absent; SR: present; SR characteristics: ellipsoidal to spheroidal membrane-bound with spheroidal granules, each ∼0.5 or the granules may be unbound and scattered among SZ; SZ: sausage-shaped, slightly larger at one end where there is a spheroidal to ellipsoidal RB; N: in center of SZ. Distinctive features of sporocyst: all sporocysts are located together at rounded end of the oocyst.

Prevalence: In 51/150 (34%) of the type host; in 5/22 (23%) *C. p. bellii* from Iowa (Wacha & Christiansen, 1976); in 1/7 (14%) *G. geographica* from Arkansas (McAllister et al., 1994); in 1/29 (3%) *T. gaigeae* from New Mexico (McAllister et al., 1995), and in 1/11 (9%) *G. caglei* from Texas (McAllister et al., 1991).

Sporulation: Likely exogenous. Turtle feces were placed into specimen jars in 2.5% aqueous (w/v) $K_2Cr_2O_7$ solution at ∼22°C for 1 week and then examined. Samples with sporulated oocysts were then stored in a refrigerator at ∼5°C.

Prepatent and patent periods: Unknown. Oocysts recovered from intestinal contents.

Site of infection: Probably the intestines because oocysts were not found in the bile by Wacha and Christiansen (1976).

Endogenous stages: Unknown.

Cross-transmission: None to date.

Pathology: Unknown.

Materials deposited: None.

Entymology: The specific epithet of this eimerian is derived from the genus name of the turtle type host.

Remarks: This species was originally described and named by Deeds and Jahn (1939) from Western painted turtles in northwestern Iowa. Wacha and Christiansen (1976) later redescribed it and added new information on the structure of the oocyst and sporocysts including the SB, PG, and "an operculum as indicated by the fracture line circumscribing the oocyst wall near the polar end." They speculated that this fracture line is the point at which the oocyst wall breaks during excystation in the next host that ingests sporulated oocysts.

Eimeria marginata (Deeds & Jahn, 1939) Pellérdy, 1974

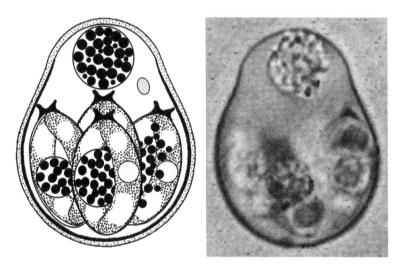

Figures 2.13, 2.14 Line drawing of the sporulated oocyst of Eimeria marginata *(original) modified from Deeds and Jahn (1939) and from Wacha and Christiansen (1976). Photomicrograph of a sporulated oocyst of* E. marginata *from Wacha and Christiansen (1976), with permission from John Wiley & Sons, Ltd. holder of the copyright for the* Journal of Eukaryotic Microbiology *(formerly* Journal of Protozoology*).*

Synonyms: *Eimeria delagei* var. *marginata* Deeds & Jahn, 1939; *Eimeria delagei marginata* Wacha & Christiansen, 1976.

Type host: *Chrysemys picta belli* (Gray, 1831) (syn. *Chrysemys bellii marginata* Conant, 1938), Western Painted Turtle.

Type locality: NORTH AMERICA: USA: Iowa, Lake West Okoboji.

Other hosts: *Graptemys geographica* (Le Sueur, 1817), Common Map Turtle; *Graptemys pseudogeographica* (Gray, 1831), False Map Turtle; *Pseudemys concinna* (LeConte, 1830), Eastern River

Cooter; *Trachemys gaigeae* (Hartweg, 1939), Big Bend Slider; *Trachemys scripta elegans* (Wied, 1838), Red Eared Slider.

Geographic distribution: NORTH AMERICA: USA: Arkansas, Iowa, New Mexico, Texas.

Description of sporulated oocyst: Oocyst shape: flask or pear-shaped; number of walls: at least two membranes (Deeds & Jahn, 1939) or single-layered, ~ 0.7 (Wacha & Christiansen, 1976); wall characteristics: thins slightly toward polar end; L × W: $20-28 \times 15-21$ (Deeds & Jahn, 1939) or L × W: 22.1×17.6 $(18.5-25 \times 16-20)$ with L/W ratio: 1.3 $(1.1-1.5)$ (Wacha & Christiansen, 1976); M: absent; PG: 1; OR: present; OR characteristics: spheroidal membrane-bound body, $3.3-9.2$ with numerous spheroidal granules, each ~ 0.7, and located near pointed or polar end. Distinctive features of oocyst: pear-shape with membrane-bound OR near pointed end.

Description of sporocyst and sporozoites: Sporocyst shape: spindle-shaped to broadly ellipsoidal; L × W: $10-14 \times 5-7$ (Deeds & Jahn, 1939) or L × W: 12.1×6.8 $(10-14 \times 5-8)$ (Wacha & Christiansen, 1976); L/W ratio: 1.8; SB: present with an inflorescence of two or more bristle-like filaments radiating outward from its surface; SSB, PSB: both absent; SR: present; SR characteristics: ellipsoidal to spheroidal, membrane-bound body, $3.3-5.3$, packed with spheroidal granules, each ~ 0.5, but sometimes granules may be unbound and scattered among SZ; SZ: banana-shaped (line drawings) with two RBs, one at each end with an N faintly visible in the middle of the SZ. Distinctive features of sporocyst: all four cluster together at the more rounded end.

Prevalence: In 10/150 (7%) of the type host from Iowa, and 2/22 (9%) from the same host, also in Iowa (Wacha & Christiansen, 1976); in 1/1 (100%) *C. p. dorsalis*, 1/7 (14%) *G. geographica*, and in 1/5 (20%) *P. concinna* from Arkansas; in 14/100 (14%) *T. s. elegans* from Texas (McAllister et al., 1994); in 1/2 (50%) *G. geographic* and 3/8 (37.5%) *G. pseudogeographica* both from Iowa (Wacha & Christiansen, 1976); and in 2/29 (7%) *T. gaigeae* from New Mexico (McAllister et al., 1995).

Sporulation: Likely exogenous. Turtle feces were placed into specimen jars in 2.5% aqueous (w/v) $K_2Cr_2O_7$ solution at $\sim 22°C$ for 1 week and then examined. Samples with sporulated oocysts were then stored in a refrigerator at $\sim 5°C$.

Prepatent and patent periods: Unknown. Oocysts recovered from intestinal contents.

Site of infection: Probably the intestines because oocysts were not found in the bile by Wacha and Christiansen (1976).
Endogenous stages: Unknown.
Cross-transmission: None to date.
Pathology: Unknown.
Materials deposited: None.
Remarks: This species was first named as a new variety of *E. delagei* by Deeds and Jahn (1939), but Pellérdy (1974) said that it bore a close resemblance to *E. chrysemydis*; however, since varieties are not commonly allowed/accepted within the Apicomplexa, he named it as a new species. Wacha and Christiansen (1976), apparently unaware of Pellérdy's name change, redescribed this species under the name *E. delagei marginata*, but we prefer to concur with Pellérdy that it should be a separate species until there is evidence to the contrary. Wacha and Christiansen (1976) also measured oocysts of this form collected from *G. pseudogeographica* that were 20.0 × 16.1 (18.5−24 × 13−18.5) with an L/W ratio of 1.2 (1.1−1.4) and sporocysts that were 11.7 × 6.3 (9−14 × 5−7).

Eimeria tetradacrutata Wacha & Christiansen, 1976

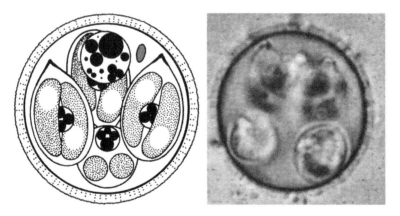

Figures 2.15, 2.16 Line drawing of the sporulated oocyst of Eimeria tetradacrutata *(original) modified from Wacha and Christiansen (1976). Photomicrograph of a sporulated oocyst of* E. tetradacrutata *from Wacha and Christiansen (1976), with permission from John Wiley & Sons, Ltd. holder of the copyright for the* Journal of Eukaryotic Microbiology *(formerly* Journal of Protozoology*).*

Type host: *Chrysemys picta belli* (Gray, 1831) (syn. *Chrysemys bellii marginata* Conant, 1938), Western Painted Turtle.
Type locality: NORTH AMERICA: USA: Iowa.

Other hosts: *Graptemys geographica* (Le Sueur, 1817), Common Map Turtle; *Trachemys scripta elegans* (Wied, 1838), Red Eared Slider.

Geographic distribution: NORTH AMERICA: USA: Arkansas, Iowa, Texas.

Description of sporulated oocyst: Oocyst shape: spheroidal to subspheroidal; number of walls: 1; wall characteristics: 1.3−1.9 thick with outer surface mammillated and appears striated in optical cross section or stippled in tangential section; L × W (N = 30): 19.5 × 19.2 (17−23 × 16−22); L × W: 1.0 (1.0−1.1); M: absent; PG: 1; OR: present; OR characteristics: spheroidal to ellipsoidal membrane-bound body, 3.8−7.7, containing several spheroidal granules, each ∼0.3−3.2. Distinctive features of oocyst: spheroidal shape with mammillated outer surface that appears striated in cross section.

Description of sporocyst and sporozoites: Sporocyst shape: ovoidal to tear drop shaped with one pointy end; L × W: 10.7 × 6.4 (9−11.5 × 6−7); L/W ratio: 1.7; SB: present as a thickened portion of the pointed tip of sporocyst; SSB, PSB: both absent; SR: present; SR characteristics: spheroidal membrane-bound body, 2.6−3.8, with a few spheroidal granules inside, each ∼0.5−2; SZ: sausage-shaped (line drawings) with one spheroidal to ellipsoidal RB at one end of SZ. Distinctive features of sporocyst: tear-drop-shaped sporocyst with a distinctly pointed end.

Prevalence: In 1/22 (4.5%) of the type host; in 3/7 (43%) *G. geographica* from Arkansas and in 3/100 (3%) *T. s. elegans* from Texas (McAllister et al., 1994).

Sporulation: Likely exogenous. Fecal material was placed in 2.5% aqueous (w/v) $K_2Cr_2O_7$ solution at ∼22°C for 1 week and then stored in a refrigerator at ∼5°C.

Prepatent and patent periods: Unknown.

Site of infection: Possibly the liver because oocysts were found only in the bile, but not in the feces of the infected type host (Wacha & Christiansen, 1976).

Endogenous stages: Unknown.

Cross-transmission: None to date.

Pathology: Unknown.

Materials deposited: None.

Entymology: The specific epithet of this eimerian is derived from the Greek to mean "four tears" referring to the four tear-drop-shaped sporocysts characteristic of this species' oocysts.

Remarks: At least two other eimerians have oocysts/sporocysts that resemble this species, *Eimeria pseudemydis* and *Eimeria trionyxae*. All three eimerian species are mostly spheroidal in shape, have an OR, and possess tear-drop-shaped sporocysts, but lack an M and conical projections on their oocysts walls. The oocysts of *E. tetradacrutata* differ from those of *E. pseudemydis* by having an L/W ratio closer to 1.0 (19.5 × 19.2, L/W 1.01 vs. 19.5 × 17.5, L/W 1.11), thicker walls which are sculptured (mammillated), a thin SB, and a PG. They differ from those of *E. trionyxae* by having larger oocysts (17−23 × 16−22 vs.14−18.5), a sculptured wall, and a PG.

Genus *Clemmys* Ritgen, 1828 (Monospecific)
To our knowledge, there are no coccidia described from this genus.

Genus *Deirochelys* Latreille, 1801 (Monospecific)
To our knowledge, there are no coccidia described from this genus.

Genus *Emydoidea* Holbrook, 1838 (Monospecific)
To our knowledge, there are no coccidia described from this genus.

Genus *Emys* Duméril,1805 (3 Species)
Eimeria delagei (Labbé, 1893) Reichenow, 1921

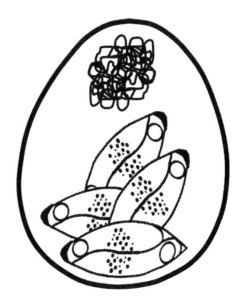

Figure 2.17 Line drawing of the sporulated oocyst of Eimeria delagei *(original) modified from* Ovezmukhammedov (1978).

Synonym: *Coccidium delagei* Labbé, 1893.

Type host: *Emys orbicularis* (L., 1758), European Pond Turtle.

Type locality: WESTERN EUROPE: France: Paris, in the laboratory at the Sorbonne University (?).

Other hosts: None to date.

Geographic distribution: WESTERN EUROPE: France.

Description of sporulated oocyst: Oocyst shape: ovoidal to ellipsoidal; number of walls: one; wall characteristics: very thin; L × W: 20 × 16–17; L/W ratio: 1.2; M, PG: both absent; OR: present; OR characteristics: visible as a pale, vacuolar structure at the narrower end of the oocyst. Distinctive features of oocyst: very thin, single-layered wall and an OR that is a pale vacuole.

Description of sporocyst and sporozoites: Sporocyst shape: unknown; L × W: unknown; L/W ratio: unknown; SB, SSB, PSB, SR: all unknown; SZ: spindle-shaped, ~7–8 long, with an N closer to the pointed end than the blunt end. Distinctive features of sporocyst: unknown.

Prevalence: In 1/1 (100%) of the type host.

Sporulation: Exogenous; sporulation takes 2–3 days (Pellérdy, 1974).

Prepatent and patent periods: Unknown.

Site of infection: Unknown.

Endogenous stages: Unknown.

Cross-transmission: None to date.

Pathology: Unknown.

Materials deposited: None.

Remarks: Labbé (1893) first reported this form as *Coccidium delagei* from *E. orbicularis* in the intestine of one captive turtle in a laboratory at the Sorbonne University (presumably the type locality). Reichenow (1921) corrected the genus name and redescribed it. Ovezmukhammedov (1978) reported finding this species in 45/125 (36%) *E. orbicularis* collected from two localities in western Turkmenistan and said that younger turtles were more heavily infected than mature ones.

Eimeria emydis Segade, Crespo, Ayres, Cordero, Arias, García−Estévez, Iglesias, & Blanco, 2006

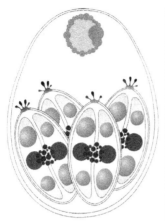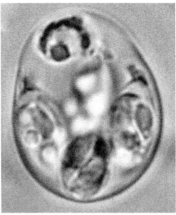

Figures 2.18, 2.19 Line drawing of the sporulated oocyst of Eimeria emydis *from Segade et al. (2006), with permission from the* Journal of Parasitology. *Photomicrograph of a sporulated oocyst of* E. emydis *from Segade et al. (2006), with permission from the* Journal of Parasitology.

Type host: *Emys orbicularis* (L., 1758), European Pond Turtle.
Type locality: WESTERN EUROPE: Spain: Galicia, Gándaras de Budiño (Pontevedra).
Other hosts: None to date.
Geographic distribution: WESTERN EUROPE: Spain.
Description of sporulated oocyst: Oocyst shape: ovoidal, rarely pear-shaped; number of walls: 1; wall characteristics: smooth, ∼0.8 thick, slightly thinner at pointed end; L × W: 22.6 × 17.0 (20−25 × 15.5−18); L/W ratio: 1.3 (1.2−1.5); M, PG: both absent; OR: present; OR characteristics: present at slightly pointed end of oocyst, ∼4.5 × 4 (4−6 × 3.5−4), consisting of an aggregate of numerous granules usually surrounding a large, clear vacuolated area. Distinctive features of oocyst: the OR is usually at the slightly pointed end of the oocyst, opposite the sporocysts, which congregate at the more rounded end.
Description of sporocyst and sporozoites: Sporocyst shape: ellipsoidal; L × W: 11.4 × 6.0 (9−13 × 5−7); L/W ratio: 1.9 (1.6−2.2); SB: present, prominent, bearing 3−5 club-shaped projections; SSB, PSB: both absent; SR: present; SR characteristics: a loose cluster of small granules (rarely consisting of scattered granules); SZ: elongate, arranged head-to-tail, each with a large spheroidal or ellipsoidal RB at broad end and a smaller RB at narrow end; N: visible between RBs in middle of SZ. Distinctive features of sporocyst: prominent SB, each with 3−5 club-shaped projections.

Prevalence: In 4/30 (13%) of the type host from the type locality; in 1/14 (7%) from the same host species from the Arnoia River (Ourense).

Sporulation: Exogenous; sporulation time not recorded.

Prepatent and patent periods: Unknown.

Site of infection: Unknown. Oocysts recovered from fecal material.

Endogenous stages: Unknown.

Cross-transmission: None to date.

Pathology: Unknown.

Materials deposited: Photosyntypes of sporulated oocysts are in the Museo Nacional de Ciencias Naturales (CSIC), Madrid, Spain, MNCN 35.01/12.

Entymology: The specific epithet is derived from the generic name of the host.

Remarks: The sporulated oocysts of this species are similar to those of *Eimeria gallaeciaensis* (below) and those of *E. marginata*, from the western painted and other turtle species. However, its ovoidal oocysts are larger than those of *E. gallaeciaensis* (22.6×17.0 [$20-25 \times 15.5-18$] vs. 19.3×16.0 [$17-22 \times 15-18$]), they have a thinner wall at the pointed end, they also have larger sporocysts (11.4×6.0 [$9-13 \times 5-7$] vs. 9.7×5.1 [$9-10 \times 5-6$]), each with $3-5$ club-shaped projections from the SB vs. the short, thin projections from the SB on the sporocysts of the latter. The oocysts of *E. marginata* are generally flask-shaped or pear-shaped (vs. ovoidal) and have sporocysts with SBs that possess two or more bristle-like (vs. club-shaped) projections.

Eimeria gallaeciaensis Segade, Crespo, Ayres, Cordero, Arias, García–Estévez, Iglesias, & Blanco, 2006

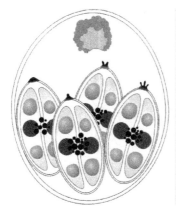 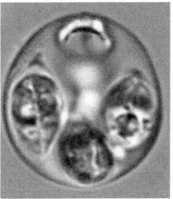

Figures 2.20, 2.21 Line drawing of the sporulated oocyst of Eimeria gallaeciaensis *from Segade et al. (2006), with permission from the* Journal of Parasitology. *Photomicrograph of a sporulated oocyst of* E. gallaeciaensis *from Segade et al. (2006), with permission from the* Journal of Parasitology.

Type host: *Emys orbicularis* (L., 1758), European Pond Turtle.

Type locality: WESTERN EUROPE: Spain: Galicia, Gándaras de Budiño (Pontevedra).

Other hosts: None to date.

Geographic distribution: WESTERN EUROPE: Spain.

Description of sporulated oocyst: Oocyst shape: subspheroidal to ovoidal−ellipsoidal; number of walls: 1; wall characteristics: smooth, ∼0.7 thick; L × W: 19.3 × 16.0 (17−22 × 15−18); L/W ratio: 1.2 (1.1−1.3); M, PG: both absent; OR: present; OR characteristics: present at end of oocyst opposite sporocysts, ∼3.8 × 3.0 (3.5−4 × 3), consisting of an irregular cluster of unbounded granules usually surrounding a vacuolar-like structure. Distinctive features of oocyst: the OR is usually at the end of the oocyst, opposite the sporocysts, which congregate at the other end.

Description of sporocyst and sporozoites: Sporocyst shape: ellipsoidal; L × W: 9.7 × 5.1 (9−10 × 5−6); L/W ratio: 1.9 (1.7−2.0); SB: present, a conical body bearing 1−4 short, thin projections; SSB, PSB: both absent; SR: present; SR characteristics: a loose cluster of small granules (rarely consisting of scattered granules); SZ: elongated sausage-shaped, arranged head-to-tail, each with a large, ellipsoidal RB at one end and another, smaller RB at the other end; N: sometimes visible in middle of SZ. Distinctive features of sporocyst: a conical SB that has 1−4 short, thin projections.

Prevalence: In 15/30 (50%) of the type host from the type locality; in 5/14 (36%) of the same host species from the Arnoia River (Ourense).

Sporulation: Exogenous; sporulation time not recorded.

Prepatent and patent periods: Unknown.

Site of infection: Unknown. Oocysts recovered from fecal material.

Endogenous stages: Unknown.

Cross-transmission: None to date.

Pathology: Unknown.

Materials deposited: Photosyntypes of sporulated oocysts are in the Museo Nacional de Ciencias Naturales (CSIC), Madrid, Spain, MNCN 35.01/11.

Entymology: The specific epithet is formed by adding the Latin termination −ensis (from) to *Gallaecia*, the Latin name of the Spanish region (Galicia) where the host species was collected.

Remarks: Sporulated oocysts of this species somewhat resemble those of *E. delagei*, first described from the European pond turtle in France (Labbé, 1893) and later from turtles in Turkmenistan (Ovezmukhammedov, 1978). Its oocysts differ from those of *E. delagei* in shape (subspheroidal vs. clearly ovoidal or ovoidal-pear-shaped; see Labbé, 1893 and McAllister & Upton, 1989) as well as the presence of short projections from the SB, which the SB of *E. delagei* sporocysts do not have.

Genus *Glyptemys* Agassiz, 1857 (2 Species)
Eimeria lecontei Upton, McAllister, & Garrett, 1995

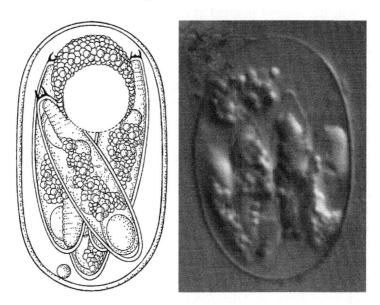

Figures 2.22, 2.23 Line drawing of the sporulated oocyst of Eimeria lecontei *from Upton et al. (1995), with permission from* Acta Protozoologica *and from C.T. McAllister. Photomicrograph of a sporulated oocyst of* E. lecontei *from Upton et al. (2006), with permission from* Acta Protozoologica *and from C.T. McAllister.*

Type host: *Glyptemys insculpta* (LeConte, 1830) (syn. *Clemmys insculpta* Fitzinger, 1835), Wood Turtle.
Type locality: NORTH AMERICA: USA: Texas, Dallas County, Dallas Zoo.
Other hosts: None to date.
Geographic distribution: NORTH AMERICA: USA: Texas.
Description of sporulated oocyst: Oocyst shape: ellipsoidal; number of walls: 1; wall characteristics: thin, smooth, delicate; L × W (N = 25):

28.3 × 15.8 (26−32 × 14−18); L/W ratio: 1.8 (1.5−2.1); M: absent; OR: present; OR characteristics: either scattered or a subspheroidal body composed of coarse granules surrounding a vacuolar-like structure, 10.5 × 8.3 (8−11 × 5−11), PG: present. Distinctive features of oocyst: presence of both a PG and a large OR that is composed of both a big vacuole surrounded by many coarse granules.

Description of sporocyst and sporozoites: Sporocyst shape: elongate-ellipsoidal, slightly pointed at one end; L × W (N = 20): 19.7 × 5.6 (18−23 × 5−6); L/W ratio: 3.6 (3−4); SB: present with blunt, short filaments; SSB, PSB: both absent; SR: present; SR characteristics: numerous coarse granules of various sizes scattered between SZ; SZ: elongate, slightly pointed at one end, 15.4 × 2.9 (14−16 × 2−3) *in situ* with various transverse striations in anterior end and a spheroidal, anterior RB, ∼2.3 (2−3) wide and a spheroidal to ellipsoidal posterior RB, 3.4 × 2.7 (2−6 × 2−3); N visible between RBs. Distinctive features of sporocyst: elongate-ellipsoidal shape with a large L/W ratio (3.6) and blunt, short filaments extending from SB.

Prevalence: In 4/4 (100%) of the type host.

Sporulation: Exogenous. Oocysts were passed only partially sporulated and completed their development within 5 days at 23°C.

Prepatent and patent periods: Unknown.

Site of infection: Unknown.

Endogenous stages: Unknown.

Cross-transmission: None to date.

Pathology: Unknown.

Materials deposited: Phototypes of sporulated oocysts are deposited in the USNPC No. 84157. A symbiotype host is deposited in the University of Texas at Arlington vertebrate collection as No. 938192.

Entymology: The specific epithet of this eimerian was given in honor of Major John Eatton LeConte (1784−1860), herpetologist, who described the type host in 1830.

Remarks: There are a number of species of turtle eimerians reported in the literature with filaments arising from the SBs on their sporocysts, but none are like those found in this species. For example, the filaments reported on the SBs of *Eimeria trachemydis* (from *T. s. elegans*), *E. filamentifera* (from *C. serpentina*) and from *Eimeria pallidus* and *Eimeria spinifera* (from *Apalone spinifera*) all are more elongate and more filamentous, whereas those on the SB of *Eimeria somervellensis* from *Pseudemys texana* are a singular, stout structure resembling the Die Pickelhaube, a spike on German

World War I helmets, and those on the SB of *Eimeria cooteri* (from *P. texana*) are also stout, covered by a thin membrane, and possess knob-like structures at their ends.

Eimeria megalostiedae Wacha & Christiansen, 1974

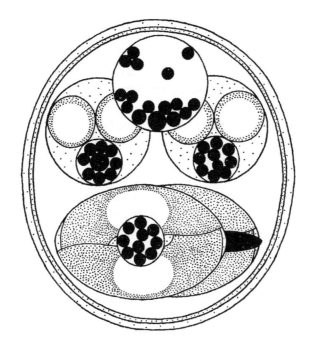

Figure 2.24 Line drawing of the sporulated oocyst of Eimeria megalostiedae *from Wacha and Christiansen (1974), with permission from* Comparative Parasitology *(formerly* Journal of the Helminthological Society of Washington*).*

Type host: *Glyptemys insculpta* (LeConte, 1830) (syn. *Clemmys insculpta* Fitzinger, 1835), Wood Turtle.
Type locality: NORTH AMERICA: USA: Iowa, Floyd County, Floyd Township, Idewild Access on Cedar River, Northeast Quarter, Section 5.
Other hosts: None to date.
Geographic distribution: NORTH AMERICA: USA: Iowa.
Description of sporulated oocyst: Oocyst shape: subspheroidal to broadly ovoidal; number of walls: 2; wall characteristics: outer layer smooth, while the inner is thinner and darker than outer; ~0.5 thick; L × W (N = 30): 13.9 × 12.8 (12–16 × 10–15); L/W ratio: 1.1 (1.0–1.2); M, PG: both absent; OR: present; OR characteristics:

spheroidal or ellipsoidal, membrane-bound body, 4−9, with several scattered spherical granules, ∼0.5−1.5 wide. Distinctive features of oocyst: walls of sporulated oocysts wrinkle readily in Sheather's solution and have a membrane-bound OR.

Description of sporocyst and sporozoites: Sporocyst shape: broadly ellipsoidal; L × W (N = 30): 9.4 × 5.1 (8−10 × 4−6); L/W ratio: 1.8; SB: present, large, ∼1.5−2 wide; SSB, PSB: both absent; SR: present; SR characteristics: spheroidal, membrane-bound body, ∼2.5−3.0 wide, with densely packed spheroidal granules, each ∼0.5−1.0 wide; SZ: one ellipsoidal RB ∼2 × 3 in each SZ. Distinctive features of sporocyst: large, flat SB on each, but which is barely visible (line drawing).

Prevalence: In 1/1 (100%) from the type host.

Sporulation: Probably exogenous, but unable to determine from the sporulation method used. Intestinal contents were removed from the turtle within 1 hr after the death of the animal. They were placed in a thin layer of 2.5% (w/v) aqueous $K_2Cr_2O_7$ solution at room temperature for 5 days to facilitate sporulation and then stored in a refrigerator until examined for the presence of oocysts.

Prepatent and patent periods: Unknown.

Site of infection: Unknown.

Endogenous stages: Unknown.

Cross-transmission: None to date.

Pathology: Unknown.

Materials deposited: None.

Entymology: The specific epithet of this eimerian is derived from a Greek derivation and means "large Stieda body."

Remarks: Wood turtles are extremely rare in Iowa, and the specimen examined from Floyd County was the first wood turtle collected in the previous 30 years (Bailey, 1941; Wacha & Christiansen, 1974). Prior to the description of this species, only about 17 *Eimeria* species from turtles were known. Wacha and Christiansen (1974) believed that the sporulated oocysts of this species could be distinguished morphologically from those of the other species of *Eimeria* reported from turtles by the following combination of characters: oocyst shape, subspheroidal to broadly ovoidal, its small size and L/W (1.1), absence of both an M and OR, presence of both OR and SR; sporocyst shape, broadly ellipsoidal, and presence of an unusually large SB (although this was not obvious in their line drawing, which showed virtually no SB).

Genus *Graptemys* Agassiz, 1857 (13 Species)
Eimeria graptemydos Wacha & Christiansen, 1976

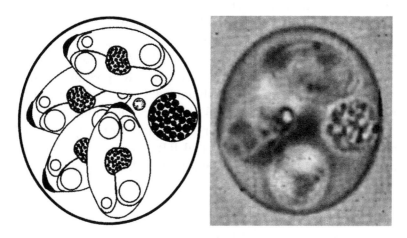

Figures 2.25, 2.26 Line drawing of the sporulated oocyst of Eimeria graptemydos *(original) adopted from Wacha and Christiansen (1976). Photomicrograph of a sporulated oocyst of* E. graptemydos *from Wacha and Christiansen (1976), with permission from John Wiley & Sons, Ltd. holder of the copyright for the* Journal of Eukaryotic Microbiology *(formerly* Journal of Protozoology*).*

Type host: *Graptemys geographica* (Le Sueur, 1817), Common Map Turtle.

Type locality: NORTH AMERICA: USA: Iowa, Gray.

Other hosts: *Chrysemys picta belli* (Gray, 1831), Western Painted Turtle; *Chrysemys picta dorsalis* Agassiz, 1857, Southern Painted Turtle; *Graptemys caglei* Haynes & McKnown, 1974, Cagle's Map Turtle; *Graptemys geographica* (Le Sueur, 1817), Common Map Turtle; *Graptemys pseudographica kohnii* (Baur, 1890), Mississippi Map Turtle; *Graptemys versa* Stejneger, 1925, Texas Map Turtle; *Kinosternon flavescens flavescens* Wermuth & Mertens, 1927, Yellow Mud Turtle; *Kinosternon subrubrum hippocrepis* Gray, 1855, Mississippi Mud Turtle; *Trachemys gaigeae* (Hartweg, 1939), Big Bend Slider; *Trachemys scripta elegans* (Wied, 1838), Red Eared Slider.

Geographic distribution: NORTH AMERICA: USA: Arkansas, Iowa, New Mexico, Texas.

Description of sporulated oocyst: Oocyst shape: broadly ellipsoidal to subspheroidal, occasionally ovoidal; number of walls: 1; wall characteristics: outer surface smooth, 0.5 thick; L × W: 12.6 × 11.4 (10−16 × 9−14.5); L/W ratio: 1.1 (1.1−1.2); M: absent;

OR: present; OR characteristics: a spheroidal, membrane-bounded, 2.6–5.3, with numerous spheroidal granules ranging in size from 0.3 to 2; PG: present. Distinctive features of oocyst: presence of both a membrane-bound OR and a PG in same oocyst.

Description of sporocyst and sporozoites: Sporocyst shape: broadly ellipsoidal; L × W: 7.3 × 4.4 (7–8 × 3–5); L/W ratio: 1.7; SB: present, thin, convex; SSB, PSB: both absent; SR: present; SR characteristics: spheroidal membrane-bounded body, 2.0–2.6, with few spheroidal granules, each ~0.3–1, or granules may be scattered within sporocyst; SZ: with a large spheroidal or ellipsoidal RB at broad end and a smaller spheroidal RB at narrow end. Distinctive features of sporocyst: presence of a small SB, a membrane-bounded SR, and two RBs in each SZ.

Prevalence: In 1/2 (50%) of the type host; in 1/1 (100%) *C. p. dorsalis*, 1/4 (25%) *G. p. kohnii*, 1/7 (14%) *G. geographica*, and in 2/6 (33%) *K. s. hippocrepis* from Arkansas, and in 10/16 (63%) *G. caglei*, 1/3 (33%) *G. versa*, 2/5 (40%) *K. f. flavescens*, and 34/100 (34%) *T. s. elegans* from Texas (McAllister et al., 1991, 1994); in 3/29 (10%) *T. gaigeae* from New Mexico (McAllister et al., 1995); in 5/8 (62.5%) *G. pseudogeographica*, and 10/22 (45%) *C. p. belli*, both from Iowa (Wacha & Christiansen, 1976); and in 1/1 (100%) *C. p. belli* from New Mexico (McAllister et al., 1995).

Sporulation: Likely exogenous. Turtle feces were placed into specimen jars in 2.5% aqueous (w/v) $K_2Cr_2O_7$ solution at ~22°C for 1 week and then examined. Samples with sporulated oocysts were then stored in a refrigerator at ~5°C.

Prepatent and patent periods: Unknown.

Site of infection: Likely the intestines because oocysts were found only in the feces from the intestine and not in the bile.

Endogenous stages: Unknown.

Cross-transmission: None to date.

Pathology: Unknown.

Materials deposited: None.

Entymology: The specific epithet of this eimerian is derived from the genus name of the turtle type host.

Remarks: The oocysts which were isolated from *G. pseudogeographica* and *C. picta belli* by Wacha and Christiansen (1976) were morphologically indistinguishable from those found in the type host and, therefore, they were considered to be the same species. Oocysts from *G. pseudogeographica* were 14.7 × 13.5 (12–18.5 × 11–17),

L/W ratio 1.1 (1.0−1.2), and sporocysts were 8.7×5.2 (7−11×4−7); oocysts from *C. picta belli* were 13.5×12.8 (11.5−16×11−16), L/W ratio 1.1 (1.0−1.1), and sporocysts were 9.6×4.5 (8−11×4−5). Several eimerians from turtles have sporulated oocysts which, like those of *E. graptemydos*, have an OR, are broadly ellipsoidal to subspheroidal in shape, lack an M, lack conical projections of the oocyst wall, and have ellipsoidal sporocysts. Sporulated oocysts of *E. graptemydos* differs from those of *E. carri* by having shorter sporocysts and a polar granule; from those of *Eimeria dericksoni* by having a SB which is thicker and more convex, and a larger, more conspicuous SR; from those of *E. lutotestudinis* by having sporocysts that are broadly ellipsoidal, an SB which is more thinly convex and less conical or pointed, and a PG; and from those of *E. megalostiedai* by having an SB which is more thinly convex and less conical or pointed, and a PG.

Eimeria juniataensis Pluto & Rothenbacher, 1976

Figure 2.27 Line drawing of the sporulated oocyst of Eimeria juniataensis *from Pluto and Rothenbacher (1976), with permission from the* Journal of Parasitology.

Type host: *Graptemys geographica* (Le Sueur, 1817), Common or Northern Map Turtle.

Type locality: NORTH AMERICA: USA: Pennsylvania, Huntindon County, Juniata River, Raystown Branch.

Other hosts: None to date.

Geographic distribution: NORTH AMERICA: USA: Pennsylvania.

Description of sporulated oocyst: Oocyst shape: spheroidal to subspheroidal (rarely ellipsoidal); number of walls: 1; wall characteristics: smooth outer wall, ~0.5 thick; L × W (N = 50): 13.5 × 12.8 (11.5−18.5 × 11.5−16.5); L/W ratio: 1.0 (1.0−1.2); M, PG: both absent; OR: present; OR characteristics: a spheroidal, membrane-bound body, ~2−5 wide, with varying numbers of small spheroidal granules usually massed together on the outside of the OR membrane. Distinctive features of oocyst: small, (mostly) spheroidal oocyst with a spheroidal, membrane-bound OR that has tiny granules sticking to it.

Description of sporocyst and sporozoites: Sporocyst shape: broadly fusiform; L × W (N = 50): 8.3 × 5.0 (7.5−10 × 4−6.5); L/W ratio: 1.7; SB: present as a small nipple-like structure sticking above sporocyst (line drawing); SSB, PSB: both absent, but the end of the sporocyst opposite the SB comes to a dull point; SR: present; SR characteristics: composed of scattered or loosely compacted spheroidal granules; SZ: comma-shaped, lying lengthwise in sporocyst with a single, large RB near the rounded end, and sometimes a smaller RB also present. Distinctive features of sporocyst: distinct nipple-like SB that protrudes from one end, while the other end comes to a dull point.

Prevalence: In 3/3 (100%) of the type host.

Sporulation: All oocysts were sporulated after 3 days at room temperature (24−26°C), but the methods used do not allow us to know whether sporulation occurs endogenously or exogenously.

Prepatent and patent periods: Unknown.

Site of infection: Unknown.

Endogenous stages: Unknown.

Cross-transmission: None to date.

Pathology: Unknown. None was observed in the three specimens of the type host.

Materials deposited: None.

Remarks: The sporulated oocysts of this species are similar to, but slightly larger than, those of *E. graptemydos* (13.5 × 12.8 [11.5−18.5 × 11.5−16.5] vs. 12.6 × 11.4 [10−16 × 9−14.5]), and they lack a PG, which is present in *E. graptemydos*. The sporocysts of *E. juniataensis* also are slightly larger than those of *E. graptemydos* (8.3 × 5.0 [7.5−10 × 4−6.5] vs. 7.3 × 4.4 [7−8 × 3−5]), have a dull point at the end opposite the SB that those of *E. graptemydos*, and lack a membrane-bounded SR that those of *E. graptemydos* possess. The sporulated oocysts of *E. juniataensis* are much smaller than those of *E. pseudogeographica* as well as other distinct differences. Based on these taxonomic features, it is likely that this is a distinct species from *Graptemys* in North America.

Eimeria pseudogeographica Wacha & Christiansen, 1976

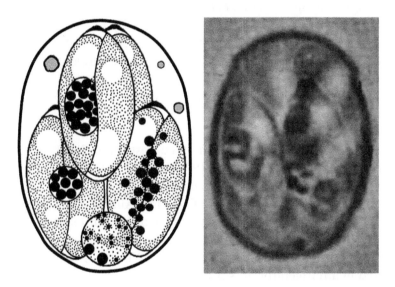

Figures 2.28, 2.29 Line drawing of the sporulated oocyst of Eimeria pseudogeographica *(original) adapted from Wacha and Christiansen (1976). Photomicrograph of a sporulated oocyst of* E. pseudogeographica *from Wacha and Christiansen (1976), with permission from John Wiley & Sons, Ltd. holder of the copyright for the* Journal of Eukaryotic Microbiology *(formerly* Journal of Protozoology*).*

Type host: *Graptemys pseudogeographica* (Gray, 1831), False Map Turtle.
Type locality: NORTH AMERICA: USA: Iowa.
Other hosts: *Chrysemys picta belli* (Gray, 1831), Western Painted Turtle; *Graptemys caglei* Haynes & McKnown, 1974, Cagle's Map Turtle; *Graptemys ouachitensis* Cagle, 1953, Ouachita Map Turtle; *Trachemys gaigeae* (Hartweg, 1939), Big Bend Slider; *Trachemys scripta elegans* (Wied, 1838), Red Eared Slider.
Geographic distribution: NORTH AMERICA: USA: Arkansas, Iowa, New Mexico, Texas.

Description of sporulated oocyst: Oocyst shape: narrowly ellipsoidal, or occasionally tapering at one end appearing slightly ovoidal; number of walls: 1; wall characteristics: outer surface smooth, ~ 0.5 thick; L $\times$ W: 19.5×13.5 $(17-21 \times 12-15)$; L/W ratio: 1.4 $(1.3-1.6)$; M: absent; OR: present; OR characteristics: spheroidal membrane-bound body, $3-5$, containing few coarse spheroidal granules, each ~ 2, scattered among numerous fine granules; PG: present as a single body, or as $2-3$ smaller randomly distributed bodies. Distinctive features of oocyst: space in this thin-walled oocyst is almost completely filled by the large sporocysts.

Description of sporocyst and sporozoites: Sporocyst shape: narrowly ellipsoidal; L $\times$ W: 11.4×5.7 $(11-13 \times 5-7)$; L/W ratio: 2.0; SB: present as a thin, convex cap on sporocyst; SSB, PSB: both absent; SR: present; SR characteristics: ellipsoidal to spheroidal membrane-bound body, $\sim 3-7$, containing numerous spheroidal granules, each ~ 1.0, or may be scattered within sporocyst; SZ: with one large ellipsoidal or spheroidal RB at broad end and smaller, spheroidal RB at narrow end.

Prevalence: In 3/8 (37.5%) of the type host; in 5/22 (23%) *C. picta belli* (Wacha & Christiansen, 1976); in 1/3 (33%) *G. ouachitensis* from Arkansas; in 1/11 (9%) *G. caglei* (McAllister et al., 1991) and 16/100 (16%) *T. s. elegans* from Texas (McAllister et al., 1994); and in 1/29 (3%) *T. gaigeae* from New Mexico (McAllister et al., 1995).

Sporulation: Likely exogenous. Fecal material was placed in 2.5% aqueous (w/v) $K_2Cr_2O_7$ solution at $\sim 22°C$ for 1 week and then stored in a refrigerator at $\sim 5°C$.

Prepatent and patent periods: Unknown.

Site of infection: Likely the intestines because oocysts of this species were found only in the feces, but not in the bile (Wacha & Christiansen, 1976).

Endogenous stages: Unknown.

Cross-transmission: None to date.

Pathology: Unknown.

Materials deposited: None.

Entymology: The specific epithet of this eimerian is derived from the specific epithet of the turtle type host.

Remarks: Wacha and Christiansen (1976) isolated oocysts from *C. p. belli* that were morphologically indistinguishable from those found in *G. pseudogeographica*, and, therefore, considered both to be those of *E. pseudogeographica*. Oocysts from *C. p. belli* were

18.9 × 14.9 (17–21 × 14–17), L/W ratio 1.3 (1.2–1.5) and sporocysts were 11.6 × 5.5 (10–13 × 4.5–6). Only *Eimeria amydae* has
sporulated oocysts, which, like those of *E. pseudogeographica*, have
an OR, are narrowly ellipsoidal to narrowly ovoidal in shape, have
ellipsoidal sporocysts, and lack both M and conical projections of
the oocyst wall. Oocysts of *E. pseudogeographica* differ from those
of *E. amydae* by having more ellipsoidal than ovoidal oocysts and
shorter sporocysts, 11.4 (11–13) vs. 14.3 (12–16).

Genus *Malaclemys* Gray, 1844 (Monospecific)
To our knowledge, there are no coccidia described from this genus.

Genus *Pseudemys* Gray, 1856 (8 Species)
Eimeria cooteri McAllister & Upton, 1989

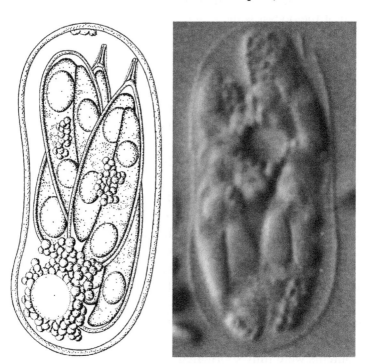

Figures 2.30, 2.31 Line drawing of the sporulated oocyst of Eimeria cooteri *from McAllister and Upton (1989), with permission from the* Canadian Journal of Zoology *and from the senior author. Photomicrograph of a sporulated oocyst of* E. cooteri *from McAllister and Upton (1989), with permission from the* Canadian Journal of Zoology *and from the senior author.*

Type host: *Pseudemys texana* Baur, 1893, Texas River Cooter.

Type locality: NORTH AMERICA: USA: Texas, Johnson County, 19.5 km SW Cleburne, off US 67 on country road 1120 at Georges Creek.

Other hosts: None to date.

Geographic distribution: NORTH AMERICA: USA: Texas.

Description of sporulated oocyst: Oocyst shape: ellipsoidal to cylindroidal, sometimes bent on one side; number of walls: 1; wall characteristics: smooth, thin, ~0.4 thick; $L \times W$ (N = 30): 25.9 × 10.9 (23−28 × 10−13); L/W ratio: 2.4 (2−3); M: absent; OR: present; OR characteristics (N = 29): a loose aggregate of granules that usually enclose a vacuolated area, 8.6 × 5.8 (6−12 × 3−10), without membranes or, rarely, scattered throughout the oocyst; PG (N = 24): a single body, ~1 wide (1−2) when not fragmented, attached to inner surface of oocyst wall. Distinctive features of oocyst: bent-cylindroidal shape with a PG that is attached to inner surface of the oocyst wall.

Description of sporocyst and sporozoites: Sporocyst shape: elongated spindle-shape, with a distinct point at one end; $L \times W$ (N = 30): 14.9 × 5.3 (13−16 × 5−7); L/W ratio: 2.8 (2−3); SB: consisting of two or more elongated structure, ~1−2 long, with tiny knob-like thickenings at the ends covered by a thin membrane; SSB, PSB: both absent; SR: present; SR characteristics (N = 30): 2.3 × 2.2 (2−4 × 2−4). In a loose aggregate, not membrane bound; SZ (N = 30): elongated, arranged head-to-tail within the sporocyst and each contains a subspheroidal or ovoidal anterior RB 2.2 × 2.3 (2−3 × 2−3), and a spheroidal posterior RB, 2.4 × 3.7 (2−3 × 3−5); N lies between the RBs. Distinctive features of sporocyst: elongated spindle-shape and the two or more unique elongated structures with tiny knob-like thickenings at the ends covered by a thin membrane that comprise what may be the SB.

Prevalence: In 3/8 (37.5%) of the type host including 2/6 (33%) in Johnson County and 1/2 (50%) in Somervell County, Texas; later found in 3/9 (33%) of the same host, also in Texas.

Sporulation: Endogenous. Oocysts recovered from feces and intestinal contents were fully sporulated.

Prepatent and patent periods: Unknown.

Site of infection: Unknown. Oocysts recovered from fecal and intestinal contents.

Endogenous stages: Unknown.

Cross-transmission: None to date.

Pathology: Unknown.

Materials deposited: A specimen of the type host is deposited in the Arkansas State University Museum of Zoology (ASUMZ 11749).

Remarks: Sporulated oocysts of this species are most similar to those of *E. trachemydis*, but they differ from it by having narrower oocysts, a larger OR, and by not possessing filaments at the ends of the SB. They also differ from those of *Eimeria texana* by its larger size, more elongate sporocysts, and lack of SB ornamentation.

Eimeria somervellensis McAllister & Upton, 1992

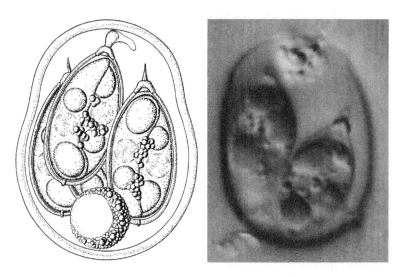

Figures 2.32, 2.33 *Line drawing of the sporulated oocyst of* Eimeria somervellensis *from McAllister and Upton (1992), with permission from the* Texas *Journal of Science and from the senior author. Photomicrograph of a sporulated oocyst of* E. somervellensis *from McAllister and Upton (1992), with permission from the* Texas Journal of Science *and from the senior author.*

Type host: *Pseudemys texana* Baur, 1893, Texas River Cooter.

Type locality: NORTH AMERICA: USA: Texas, Somervell County, 14.5 km NW Glen Rose, off county road 308 at Georges Creek.

Other hosts: *Pseudemys concinna metteri* Ward, 1984, Missouri River Cooter.

Geographic distribution: NORTH AMERICA: USA: Arkansas, Texas.

Description of sporulated oocyst: Oocyst shape: slightly pear-shaped; number of walls: 1; wall characteristics: smooth, thin, ~0.8 thick; L × W: 21.1 × 16.1 (17−23 × 14−17); L/W ratio: 1.3 (1.2−1.4); M: absent; OR: present; OR characteristics: spheroidal, ~5.9 (5−7), with a vacuole surrounded by coarse granules, PG: present (occasionally) attached either to the OR or to a sporocyst. Distinctive features of oocyst: pear-shape with a large OR of coarse granules enclosing a large vacuole.

Description of sporocyst and sporozoites: Sporocyst shape: elongate-ovoidal with one end pointed very sharply; $L \times W$: 12.8×6.5 ($11-14 \times 6-7$); L/W ratio: 2.0 (1.7–2.3); SB: elaborate and dome-shaped, resembling the spike on the German World War I helmet (*Die Pickelhaube*), 1.6×2.1 ($1-2 \times 2$); SSB: absent; PSB: may be present as a button-like process at the end of sporocyst opposite the SB; SR: present; SR characteristics (N = 20): often scattered granules, but sometimes a cluster of large granules, 3.0 (2–4) wide; SZ (N = 20): elongated banana-shaped with one end slightly thicker, 11.7×3.1, arranged head-to-tail, and each contained a spheroidal to ovoidal RB, 2.2×2.3, and a spheroidal to ovoidal posterior RB, 2.9×3.5, with a distinct N between the RBs. Distinctive features of sporocyst: elongated ovoidal shape and the elaborate SB and associated structure(s).

Prevalence: Found in 1/9 (33%) *P. texana* (includes 1/1 in original description and 0/8 *P. texana* from an earlier survey by McAllister & Upton, 1989); in 1/1 (100%) *P. concinna metteri* from Arkansas (McAllister & Upton, 1992); and in 1/5 (20%) *P. c. metteri* in Arkansas (McAllister et al., 1994).

Sporulation: Endogenous. Oocysts recovered from feces and intestinal contents were fully sporulated.

Prepatent and patent periods: Unknown.

Site of infection: Unknown.

Endogenous stages: Unknown.

Cross-transmission: None to date.

Pathology: Unknown.

Materials deposited: Voucher specimens of the type host are deposited in the Arkansas State Museum of Zoology, ASUMZ 17888. Syntypes of oocysts in 10% formalin are deposited in the US National Parasite Collection, Beltsville, Maryland, USA, as USNPC No. 80865.

Entymology: The nomen triviale combines the name of the county in Texas from which the type host was collected and -*ensis* (L., belonging to).

Remarks: Sporulated oocysts of this species most closely resemble those of *E. marginata*, but its sporocysts are less robust and the SB is more elaborate.

Eimeria texana McAllister & Upton, 1989b

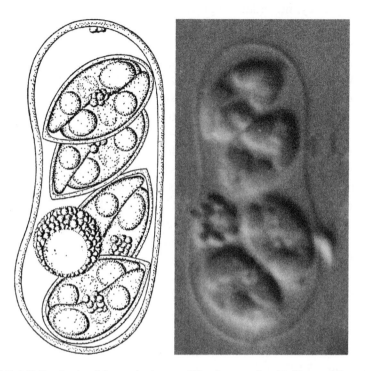

Figures 2.34, 2.35 Line drawing of the sporulated oocyst of Eimeria texana *from McAllister and Upton (1989b), with permission from the* Canadian Journal of Zoology *and from the senior author. Photomicrograph of a sporulated oocyst of* E. texana *from McAllister and Upton (1989b), with permission from the* Canadian Journal of Zoology *and from the senior author.*

Type host: *Pseudemys texana* Baur, 1893, Texas River Cooter.
Type locality: NORTH AMERICA: USA: Texas, Johnson County, 19.5 km SW Cleburne, off US 67 on county road 1120 at Georges Creek.
Other hosts: None to date.
Geographic distribution: NORTH AMERICA: USA: Texas.
Description of sporulated oocyst: Oocyst shape: elongate-cylindroidal that is sometimes indented in the middle; number of walls: 1; wall characteristics: smooth, thin, ~0.4 thick; L × W (N = 30): 20.5 × 8.4 (18−23 × 7−9); L/W ratio: 2.4 (2−3); M: absent; OR: present; OR characteristics (N = 29): more-or-less spheroidal cluster of granules that enclose a large vacuole and are bounded by a thin membrane, 4.3 × 3.9 (3−6 × 3−6) or, rarely, the granules are scattered throughout oocyst; PG: a single ellipsoidal

body attached to inner surface of oocyst wall. Distinctive features of oocyst: cylindroidal shape indented in the middle with both a membrane-bounded OR and a distinct PG attached to the inner oocyst wall.

Description of sporocyst and sporozoites: Sporocyst shape: ovoidal; L × W (N = 30): 8.1 × 4.7 (7−9 × 4−5); L/W ratio: 1.7 (1.5−2); SB: tiny, present at one end of sporocyst giving the appearance of a simple point; SSB, PSB: both absent; SR: present; SR characteristics (N = 29): 1.9 × 1.9 (1−2 × 1−2) wide, composed of a cluster of small granules that do not appear to be membrane bound; SZ (N = 30): sausage-shaped, 7.3 × 2.2 (6−8 × 2−3), arranged head-to-tail within the sporocyst, and each contained a spheroidal to ovoidal anterior RB, 1.9 × 1.8 (1−2 × 1−2) and a posterior spheroidal RB, 2 × 2.5 (2 × 2−3); a N lies between RBs. Distinctive features of sporocyst: tiny SB, distinct SR, and SZs with two RBs each.

Prevalence: In 2/8 (25%) of the type host including 1/6 (17%) from Johnson County and 1/2 (50%) in Somervell County, Texas; later found in 2/9 (22%) from the same host, also in Texas.

Sporulation: Endogenous. Oocysts recovered from feces and intestinal contents were fully sporulated.

Prepatent and patent periods: Unknown.

Site of infection: Unknown. Oocysts were recovered from fecal and intestinal contents.

Endogenous stages: Unknown.

Cross-transmission: None to date.

Pathology: None observed.

Materials deposited: Voucher specimen of the type host is deposited in the Arkansas State University Museum of Zoology (ASUMZ 11749).

Entymology: The specific epithet is derived from the scientific name of the type host.

Remarks: Sporulated oocysts of this species are most similar to those of *E. pseudogeographica*, but they may be distinguished by having a narrower oocyst, much larger L/W ratio (2.4 vs.1.4) and smaller sporocysts with a smaller SR.

Genus *Terrapene* Merrem, 1820 (4 Species)
Eimeria carri Ernst & Forrester, 1973

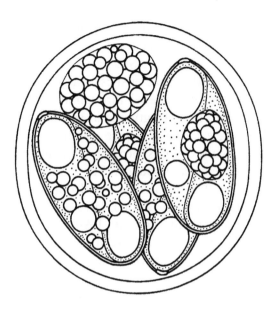

Figure 2.36 Line drawing of the sporulated oocyst of Eimeria carri *from Ernst and Forrester (1973), with permission from the* Journal of Parasitology.

Type host: *Terrapene carolina* (L., 1758), Box Turtle.

Type locality: NORTH AMERICA: USA: Alabama, Lee County.

Other hosts: *Terrapene carolina carolina* (L., 1758), Eastern Box Turtle; *Terrapene carolina triunguis* (Agassiz, 1857), Three-Toed Box Turtle; *Terrapene ornata ornata* (Agassiz, 1857), Ornate Box Turtle.

Geographic distribution: NORTH AMERICA: USA: Alabama, Arkansas, Florida.

Description of sporulated oocyst: Oocyst shape: subspheroidal to broadly ellipsoidal, rarely spheroidal; number of walls: 1; wall characteristics: smooth, colorless to light blue, ~0.7 thick; L × W (N = 50): 15.9 × 14.5 (12−20 × 12−16); L/W ratio: 1.1 (1.0−1.2); M, PG: both absent; OR: present; OR characteristics: composed of many granules of various sizes, surrounded by a thin membrane or sometimes scattered in oocyst. Distinctive features of oocyst:

spheroidal with a membrane-bounded, ellipsoidal OR of many large granules, ~6–7 long (extrapolated from line drawing; actual measurements not given).

Description of sporocyst and sporozoites: Sporocyst shape: elongate-ellipsoidal, very slightly pointed at one end; L × W: 11.1 × 5.2 (10–13 × 4–6); L/W ratio: 2.1; SB: present, tiny; SSB, PSB: both absent; SR: present; SR characteristics: many granules, surrounded by a thin membrane or scattered throughout sporocyst; SZ: elongate, banana-shaped, lying lengthwise in sporocyst, partially curled around each other, each with a large posterior and small anterior RB. Distinctive features of sporocyst: elongate-ellipsoidal shape and containing a membrane-bounded SR.

Prevalence: In 1/5 (20%) of the type host from Alabama, and in 1/2 (50%) of the same host from Florida; later, in 3/9 (33%) *T. c. triunguis* from Arkansas (McAllister et al., 1994).

Sporulation: Unknown. Oocysts were sporulated after 1 wk in 2.5% aqueous potassium dichromate ($K_2Cr_2O_7$) solution at room temperature (20–24°C).

Prepatent and patent periods: Unknown.

Site of infection: Unknown. Oocysts recovered from intestinal contents.

Endogenous stages: Unknown.

Cross-transmission: None to date.

Pathology: Unknown.

Materials deposited: None.

Remarks: This was the first eimerian described from turtles in the genus *Terrapene* and McAllister and Upton (1989a) described the only other one. The sporulated oocysts of the two species are reasonably similar in size and shape (15.9 × 14.5 [12–20 × 12–16], L/W 1.1 vs. 17.9 × 15.7 [16–21 × 14–18], L/W 1.1), as are their sporocysts, and both species possess an OR, SR, and SB. They were described from different *Terrapene* species from widely separated coasts of the USA and McAllister and Upton (1989) presented an argument that the description of their new coccidium (below) should be based primarily on marked differences in sporulation. We are inclined to believe that these two forms may comprise the same species since host specificity seems (from the data available) to be reasonably lax in turtles, but we must wait for molecular evidence to provide the final decision.

Eimeria ornata McAllister & Upton, 1989a

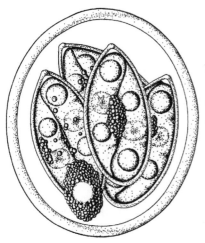

Figures 2.37, 2.38 Line drawing of the sporulated oocyst of Eimeria ornata *from McAllister and Upton (1989a), with permission from John Wiley & Sons, Ltd. holder of the copyright for the* Journal of Eukaryotic Microbiology *(formerly* Journal of Protozoology*). Photomicrograph of a sporulated oocyst of* E. ornata *from McAllister and Upton (1989a), with permission from John Wiley & Sons, Ltd. holder of the copyright for the* Journal of Eukaryotic Microbiology *(formerly* Journal of Protozoology*).*

Type host: *Terrapene ornata ornata* (Agassiz, 1857), Ornate Box Turtle.

Type locality: NORTH AMERICA: USA: Texas, Ellis County, 6.4 km SW Midlothian at Ward Spur along Soap Creek.

Other hosts: None to date.

Geographic distribution: NORTH AMERICA: USA: Texas.

Description of sporulated oocyst: Oocyst shape: ellipsoidal; number of walls: 1; wall characteristics: smooth, colorless, ~1.0 thick; L × W (N = 30): 17.9 × 15.7 (16−21 × 14−18); L/W ratio: 1.1 (1.0−1.3); M: absent; PG: sometimes present (in 10/30 oocysts measured); OR: present; OR characteristics: either as an aggregate of numerous small globules surrounding a large vacuole on one side of the OR, but sometimes as diffuse granules scattered in oocyst. Distinctive features of oocyst: OR of many smaller globules surrounding a large vacuole and presence of a PG in about 30% of the oocysts in a sample.

Description of sporocyst and sporozoites: Sporocyst shape: elongate-ellipsoidal, very slightly pointed at one end; L × W: 11.1 × 5.4 (9−13 × 5−6); L/W ratio: 2.1 (1.7−2.3); SB: present, small and indistinct; SSB, PSB: both absent; SR: present; SR characteristics: either as a compact mass of granules, 3.6×2.6 $(2-6 \times 2-4)$, or scattered as loose granules among SZ; SZ: banana-shaped, 9.5×2 $(8-12 \times 2)$, arranged head-to-tail, each with an anterior, 1.5 $(1-2)$ wide, and a posterior, 1.8 wide, RB. Distinctive features of sporocyst: elongate-ellipsoidal shape with a tiny SB, and containing a distinct SR of many granules.

Prevalence: In 6/16 (37.5%) of the type host, including 2/3 (67%) from Ellis County, 3/12 (25%) from Johnson County, and in 1/1 (100%) from Somervell County; later, it was found in 8/18 (44%) of the type host, also in Texas (McAllister et al., 1994).

Sporulation: Endogenous. Oocysts were passed fully sporulated.

Prepatent and patent periods: Unknown.

Site of infection: Unknown. Oocysts were found in the feces.

Endogenous stages: Unknown.

Cross-transmission: None to date.

Pathology: Unknown.

Material deposited: None.

Entymology: The specific epithet is derived from the scientific name of the type host.

Remarks: This species has sporulated oocysts that are somewhat similar to about a half-dozen other eimerians from turtles in different genera and families; these need not concern us here, but to evaluate the comparisons, see the *Remarks* section in McAllister and Upton (1989a). The major concern in comparing structures should be with the oocysts of *E. carri*, the only other eimerian known to date from *Terrapene* species. Although the oocysts and sporocysts of both are quite similar, there are minor structural differences. A vacuolated area was not reported present in the OR of *E. carri*, nor was a PG ever seen. Additionally, oocysts of *E. carri* were reported to sporulate exogenously after 1 week at room temperature $(20-24°C)$, while those of *E. ornata* purportedly sporulate endogenously. Finally, when we examine the historic range, ecology, and habitat requirements of the host species for these two eimerians, an adaptive explanation for the differences in sporulation seems apparent (McAllister & Upton, 1989). That is, *T. ornata* inhabits xeric areas of the south and Midwest and, thus, it is unlike its eastern cousin, *T. carolina*, which prefers more humid woodland habitats, because it can tolerate more arid conditions. Thus, McAllister and Upton (1989a) argue, "to coevolve with its host and persist in a harsher environment, it appears that it may have been necessary for *E. ornata* to be passed in the feces fully sporulated and capable of infecting another turtle immediately upon deposition, [while] *E. carri* sporulates exogenously," "...possibly due to a higher relative humidity and more precipitation in the east."

Genus *Trachemys* Agassiz, 1857 (15 Species)
Eimeria pseudemydis Lainson, 1968

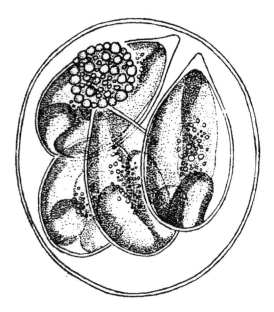

Figure 2.39 Line drawing of the sporulated oocyst of Eimeria pseudemydis *from Lainson (1968), with permission from* Pathogens & Global Health *(formerly* Annals of Tropical Medicine & Parasitology*) and from the author.*

Type host: *Trachemys ornata* (Gray, 1831) (syn. *Pseudemys ornata* Stuart, 1937), Ornate Slider.

Type locality: CENTRAL AMERICA: Belize (formerly British Honduras): El Cayo District, near Baking Pot.

Other hosts: *Deirochelys reticularia miaria* Schwartz, 1956, Western Chicken Turtle; *Glyptemys insculpta* (LeConte, 1830) (syn. *Clemmys insculpta* Fitzinger, 1835), Wood Turtle; *Pseudemys texana* Baur, 1893, Texas River Cooter; *Trachemys gaigeae* (Hartweg, 1939), Big Bend Slider; *Trachemys scripta elegans* (Wied, 1838), Red Eared Slider.

Geographic distribution: CENTRAL AMERICA: Belize; NORTH AMERICA: USA: Arkansas, New Mexico, Texas.

Description of sporulated oocyst: Oocyst shape: spheroidal (Bone, 1975) to broadly ellipsoidal (Lainson, 1968); number of walls: 1 (Lainson, 1968) or 2 (Bone, 1975); wall characteristics: very thin, smooth, colorless, ~ 1.4 thick (Bone, 1975); $L \times W$ (N = 15): 19.0×17.5 (18−20 × 16.5−18); L/W ratio: 1.1 (Lainson, 1968) or $L \times W$ (N = 50): 16.7×17.7 (16−20 × 13−19); L/W ratio: 1.1 (1.0−1.2) (Bone, 1975); M: absent; PG: absent (Lainson, 1968) or present (Bone, 1975); OR: present; OR characteristics: compact, spheroidal mass containing highly refractile granules of various sizes (Lainson, 1968) and may be bounded by a thin membrane (Bone, 1975) or occasionally scattered. Distinctive features of oocyst: very thin wall and a large, spheroidal OR that may be membrane-bounded.

Description of sporocyst and sporozoites: Sporocyst shape: pear-shaped with wall drawn out into a pointed SB at the narrower end (Lainson, 1968) or elongate-ovoidal to pyriform (Bone, 1975); $L \times W$: 12.5×6.5 (11 [sic] × 6−7); L/W ratio: 1.9 (Lainson, 1968) or 10.4×5.4 (9−13 × 4−7); L/W ratio: 1.9 (1.4−2.6) (Bone, 1975); SB: present as a distinct point at one end of sporocyst; SSB, PSB: both absent; SR: present; SR characteristics: loosely aggregated and composed of relatively fine granules (Lainson, 1968) or as numerous granules in the center of the sporocyst, and surrounded by a membrane; SZ: sausage-shaped and recurved back upon themselves with no RBs or N visible (Lainson, 1968) or vermiform, generally not intertwined or overlapping and the spheroidal RB, ~ 3 wide, at their broad end, with a smaller RB occasionally present at the more pointed end (Bone, 1975); N, ~ 1.5 wide. Distinctive features of sporocyst: the drawn-out end with the SB that gives a pointed appearance.

Prevalence: In 2/2 (100%) of the type host; in 3/4 (75%) *G. insculpta* (syn. *C. insculpta*) from Texas (McAllister et al., 1994); in 1/1 (100%) *D. r. miaria*, 2/9 (22%) *P. texana*, and 12/100 (12%) *T. s. elegans* from Texas (McAllister et al., 1994); in 1/8 (12.5%) *T. s. elegans* from Arkansas (Bone, 1975); and in 1/29 (3%) *T. gaigeae* from New Mexico (McAllister et al., 1995).

Sporulation: Unknown.

Prepatent and patent periods: Unknown.

Site of infection: Unknown, although Lainson (1968) said it was "probably the intestine."

Endogenous stages: Unknown.

Cross-transmission: None to date.

Pathology: Unknown.

Materials deposited: None.

Remarks: The only eimerian at the time with oocysts that approximated those of the form described by Lainson (1968) was *E. dericksoni* from *Apalone spinifera* (syn. *Amyda spinifera*). The oocysts of *E. dericksoni* are 14.6 × 12.9 (12−17 × 11−16), while those of *E. pseudemydis* are slightly larger (19.0 × 17.5 [18−20 × 16.5−18]), while the sporocysts of the former are only ∼7 long, while their width was not stated versus sporocysts that are about 12 (?) long in this species and the sporocysts of *E. dericksoni* do not possess the strikingly pointed SB described by Lainson (1968). The SR of *E. dericksoni* is a compact, spheroidal mass, unlike the scattered granules in *E. pseudemydis*.

Eimeria scriptae Sampson & Ernst, 1969

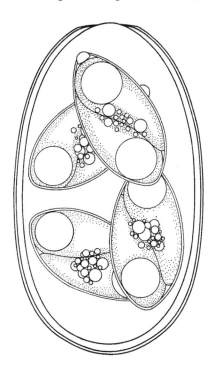

Figure 2.40 Line drawing of the sporulated oocyst of Eimeria scriptae *from Sampson and Ernst (1969), with permission from John Wiley & Sons, Ltd. holder of the copyright for the* Journal of Eukaryotic Microbiology *(formerly* Journal of Protozoology*).*

Type host: *Trachemys scripta elegans* (Wied, 1838) (syn. *Pseudemys scripta elegans* Cagle, 1944), Red Eared Slider.

Type locality: NORTH AMERICA: USA: Wisconsin, a biological supply house that collects this species of turtle along the Mississippi River Drainage from Arkansas, Illinois, Kentucky, Louisiana, Minnesota, Mississippi, Missouri, Tennessee, and Wisconsin. Thus, it is impossible to establish a more exact geographic location.

Other hosts: None to date.

Geographic distribution: NORTH AMERICA: USA: Mississippi River Drainage from Arkansas, Illinois, Kentucky, Louisiana, Minnesota, Mississippi, Missouri, Tennessee, and Wisconsin.

Description of sporulated oocyst: Oocyst shape: ovoidal to broadly ellipsoidal; number of walls: 2 or 3; wall characteristics: smooth, outer wall, ~1.25 thick, bright bluish-green, while inner layer is thinner, light brown, ~0.3−0.5 thick; a third, innermost, layer may be present, but only two layers were found after crushing the oocysts by applying pressure on the coverslip; L × W (N = 52): 24.2 × 13.7 (22−27 × 12−16); L/W ratio: 1.8 (1.5−2.1); M: present, rising slightly above the normal curvature of the oocyst wall, ~1.8 wide; OR, PG: both absent. Distinctive features of oocyst: elongated-ellipsoidal shape, with 2−3 wall layers, presence of a M, and lacking both OR and PG.

Description of sporocyst and sporozoites: Sporocyst shape: ovoidal; L × W (N = 56): 11 × 5.9 (10−14 × 5−7); L/W ratio: 1.9; SB: moderate-sized at slightly pointed end of sporocyst; SSB, PSB: both absent; SR: present; SR characteristics: granules of variable numbers and sizes occur in a compact mass or scattered loosely in the central portion of the sporocyst and, rarely, the residual granules appeared to be bounded by a thin membrane; SZ: vermiform, lying side-by-side with a spheroidal RB, ~3.0 wide, at the more rounded end of each SZ. Distinctive features of sporocyst: ovoidal shape with "typical" SZ, each with one RB, and a SR of various-sized granules.

Prevalence: In 1/43 (2%) of the type host; later found in 2/100 (2%) in the same host species from Texas (McAllister et al., 1994).

Sporulation: Unknown.

Prepatent and patent periods: Unknown.

Site of infection: Unknown.

Endogenous stages: Unknown.

Cross-transmission: None to date.

Pathology: Unknown.

Materials deposited: None.

Remarks: As far as Sampson and Ernst (1969) were able to determine, this was the first report of an *Eimeria* species in the red eared slider at that time. Their form differed from all but two of the previously described eimerians, *Eimeria brodeni* and *Eimeria chrysemydis*, by the size and the shape of the oocysts and by the possession of an M. Sporulated oocysts of *E. scriptae* differed from those of *E. brodeni* and *E. chrysemydis* by having a distinct SB, which the others lack. The oocysts of *Eimeria scriptae* differ from *E. chrysemydis* by lacking an OR and by being narrower (24.2 × 13.7 vs. 23 × 15).

Eimeria stylosa McAllister & Upton, 1989b

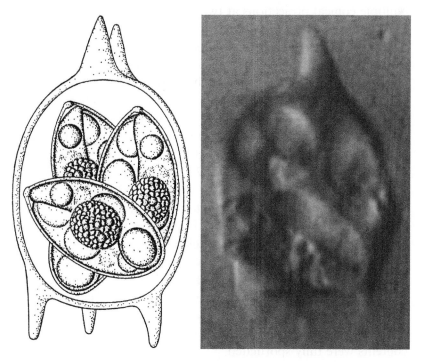

Figures 2.41, 2.42 Line drawing of the sporulated oocyst of Eimeria stylosa *from McAllister and Upton (1989b), with permission from the* Canadian Journal of Zoology *and from the senior author. Photomicrograph of a sporulated oocyst of* E. stylosa *from McAllister and Upton (1989b), with permission from the* Canadian Journal of Zoology *and from the senior author.*

Type host: *Trachemys scripta elegans* (Wied, 1838), Red Eared Slider.

Type locality: NORTH AMERICA: USA: Texas, Johnson County, 13 km SSW Cleburne, on US 67.

Other hosts: *Trachemys gaigeae* (Hartweg, 1939), Big Bend Slider.

Geographic distribution: NORTH AMERICA: USA: New Mexico, Texas.

Description of sporulated oocyst: Oocyst shape: ovoidal, with conical projections ~ 4.0 (3−5) long; most have two projections on one end and three on the opposite end of the oocyst, however, three specimens with two and two, two with four and two, and one with four and seven projections were observed; number of walls: 1; wall characteristics: smooth, colorless, ~ 0.8 thick; $L \times W$ ($N = 25$): 16.5×13.1 ($14-18 \times 12-14$); L/W ratio: 1.3 (1.2−1.5); M, OR, PG: all absent. Distinctive features of oocyst: unique shape with multiple conical projections at two ends of the oocyst.

Description of sporocyst and sporozoites: Sporocyst shape: ovoidal to slightly ellipsoidal, with a smooth, thin, colorless wall; $L \times W$ ($N = 25$): 11.1×5.8 ($10-14 \times 5-7$); L/W ratio: 1.9 (1.6−2.6); SB: present as a small knob at the slightly pointed end; SSB, PSB: both absent; SR: present; SR characteristics ($N = 19$): spheroidal, ovoidal, or (rarely) scattered, 4.0×3.4 ($3-6 \times 3-5$); SZ ($N = 25$): banana-shaped, 9.4×2.8 ($9-10 \times 2-3$), arranged head-to-tail within the sporocyst and reflected along opposite poles; each SZ with a spheroidal anterior RB, ~ 2 wide, and two larger spheroidal to subspheroidal posterior RBs, which are 2.5×3 ($2-3 \times 2-4$) with an N located between the anterior RB and the more centrally located of the two posterior RBs. Distinctive features of sporocyst: spheroidal SR and three RBs in each SZ.

Prevalence: In 2/16 (12.5%) of the type host including 2/13 (15%) from Johnson County and 0/3 (0%) from Somervell County; later, in 4/100 (4%) in the same host also from Texas (McAllister et al., 1994); in 1/29 (3%) in *T. gaigeae* from New Mexico (McAllister et al., 1995).

Sporulation: Endogenous. Oocysts recovered from feces or intestinal contents were fully sporulated.

Prepatent and patent periods: Unknown.

Site of infection: Unknown. Oocysts were recovered from fecal and intestinal contents.

Endogenous stages: Unknown.

Cross-transmission: None to date.

Pathology: Unknown.

Materials deposited: A specimen of the type host is deposited in the Arkansas State University Museum of Zoology (ASUMZ 8493).

Entymology: The specific epithet refers to the stylet-like projections on the oocyst wall.

Remarks: This is the unknown eimerian that McAllister and Upton (1988) alluded to, but did not name, because they recovered it from only 1/71 (1.4%) *T. s. elegans* in the year prior to their published description. *Eimeria stylosa* most closely resembles *E. mitraria* because of the ornate projections on the oocyst wall. However, oocysts of *E. mitraria* are smaller and possess fewer and shorter projections than this form. Wacha and Christiansen (1976) also reported oocysts of an *Eimeria* sp. in *G. geographica*, with similar morphologic characteristics to those of *E. stylosa*. They (Wacha & Christiansen, 1976) measured a single oocyst, which was slightly larger than *E. stylosa* and possessed only two projections at either end of the oocyst.

Eimeria trachemydis McAllister & Upton, 1988

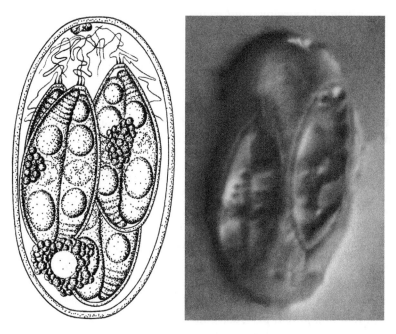

Figures 2.43, 2.44 Line drawing of the sporulated oocyst of Eimeria trachemydis *from McAllister and Upton (1988), with permission from the* Journal of Parasitology *and from the senior author. Photomicrograph of a sporulated oocyst of* E. trachemydis *from McAllister and Upton (1988), with permission from the* Journal of Parasitology *and from the senior author.*

Type host: *Trachemys scripta elegans* (Wied, 1838), Red Eared Slider.

Type locality: NORTH AMERICA: USA: Texas, Dallas County, DeSoto, 1.6 km So I−20 on Bolton Boone Drive at Terrell Pond.

Other hosts: *Chrysemys picta belli* (Gray, 1831), Western Painted Turtle; *Graptemys caglei* Haynes & McKnown, 1974, Cagle's Map Turtle; *Trachemys gaigeae* (Hartweg, 1939), Big Bend Slider.

Geographic distribution: NORTH AMERICA: USA: New Mexico, Texas.

Description of sporulated oocyst: Oocyst shape: ellipsoidal; number of walls: 1; wall characteristics: smooth, thin, ~0.6 thick; L × W (N = 18): 25.0 × 13.6 (21−30 × 12−16); L/W ratio: 1.8 (1.5−2.2); M: absent; OR: present; OR characteristics: when aggregated, residual granules usually enclosing a large vacuolated area and are 6.8 × 6.0 (4−10 × 3−10), but sometimes present only as loose granules scattered throughout oocyst; PG: present, an ellipsoidal−spheroidal body present at one pole, attached to inner surface of oocyst wall. Distinctive features of oocyst: large ellipsoidal structure, with a PG that is attached to inner surface of oocyst wall.

Description of sporocyst and sporozoites: Sporocyst shape: elongate-ovoidal; L × W: 14.4 × 5.6 (13−16 × 5−6); L/W ratio: 2.6 (2.3−2.9); SB: small, flattened, ~1.6 wide, present at slightly pointed end of sporocyst, with stalks bearing 2−5 filaments/sporocyst, each is 6−10 long; SSB, PSB: both absent; SR: present; SR characteristics: 4.0 × 3.0 (2−10 × 2−4) in a single cluster, but not membrane-bounded; SZ: arranged head-to-tail, and each contains a spheroidal or ellipsoidal anterior RB, 2.4 × 2.7 (2−3 × 2−3), and a midposterior, spheroidal RB, 2.5 × 3.5 (2−3 × 2−6); an N lies between RBs. Distinctive features of sporocyst: flattened SB with stalks bearing 2−5 filaments/sporocyst.

Prevalence: In 3/71 (4%) of the type host and 7/100 (7%) of the same host species also from Texas (McAllister et al., 1991); in 1/16 (6%) *G. caglei* from Texas (McAllister et al., 1994); in 1/29 (3%) *T. gaigeae* from New Mexico (McAllister et al., 1995); and in 1/1 (100%) *C. p. belli* from New Mexico.

Sporulation: Endogenous. Oocysts were passed fully sporulated.

Prepatent and patent periods: Unknown.

Site of infection: Unknown. Oocysts found in feces and intestinal contents.

Endogenous stages: Unknown.

Cross-transmission: None to date.

Pathology: Unknown.

Materials deposited: Voucher specimens of the type host in the Arkansas State University Museum of Zoology (ASUMZ 8475, 8493, 8553); Syntypes of sporulated oocysts in 10% formalin in the US National Parasite Museum, Beltsville, Maryland as USNPC No. 80459.

Remarks: Sporulated oocysts of this species most closely represent those of *E. filamentifera* from the common snapping turtle because of the presence of filaments at the end of the SB. However, sporocysts of the latter are wider and only slightly ovoidal to ellipsoidal, possess much shorter filaments, and lack a PG. The differences in these characteristics easily distinguish between the two species.

In addition to being infected with *E. trachemydis*, the three infected turtles studied by McAllister and Upton (1988) were multiply infected with other eimerians. These included 2/3 (67%) infected with *E. lutotestudinis*, 2/3 (67%) infected with *E. pseudogeographica*, and 1/3 (33%) infected also with *E. scriptae*. These latter three eimerians were originally described from other host genera and, if these identifications based only on morphology are correct, this suggests that little host specificity exists among certain turtle coccidia. Of course, species identifications confirmed by molecular signatures of species from different host genera are needed to corroborate this lack of host specificity.

FAMILY TESTUDINIDAE, TORTOISES, 15 GENERA, 57 SPECIES

Genus *Aldabrachelys* Loveridge and Williams, 1957 (3 Species)

To our knowledge, there are no coccidia described from this genus.

Genus *Astrochelys* Gray, 1873 (2 Species)

To our knowledge, there is an intranuclear "coccidian" described from *Astrochelys radiate* (Shaw, 1802), but it has not yet been identified to genus (Jacobson et al., 1994). In addition, Greiner (2003) showed a photomicrograph of a *Caryospora* sp., also from *A. radiate*, but did not describe or name it. See our discussion under *Species Inquirendae* (Chapter 5).

Genus *Chelonoidis* Fitzgerald, 1835 (13 Species)

Eimeria amazonensis Lainson, Da Silva, Franco, & De Souza, 2008

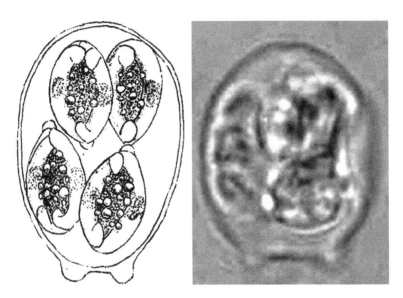

Figures 2.45, 2.46 Line drawing of the sporulated oocyst of Eimeria amazonensis *from Lainson et al. (2008), with permission from the journal* Parasite *and from the senior author. Photomicrograph of a sporulated oocyst of* E. amazonensis *from Lainson et al. (2008), with permission from the journal* Parasite *and from the senior author.*

Type host: *Chelonoidis carbonaria* (Spix, 1824) (syn. *Geochelone carbonaria* Ernst & Barbour, 1989, Red-Footed Tortoise.

Type locality: SOUTH AMERICA: (North) Brazil: Pará State, Serra dos Carajás, Salobo (6° S, 58° 18′ W).

Other hosts: Unknown.

Geographic distribution: SOUTH AMERICA: (North) Brazil: Pará State.

Description of sporulated oocyst: Oocyst shape: ovoidal; number of walls: 1; wall characteristics: a single, colorless, fragile layer, ~0.5 thick, with two short, blunt protrusions at one end; L × W (N = 50): 11.7 × 9.1 (10−13 × 8−10); L/W ratio: 1.8 (1.5−2.0); M, OR, PG: all absent. Distinctive features of oocyst: the irregular shape with two polar protrusions and no M, OR, or PG.

Description of sporocyst and sporozoites: Sporocyst shape: ellipsoidal; L × W (N = 50): 6.5 × 3.7 (5−7 × 3−4); L/W ratio: 1.7 (1.6 × 2.0); SB: present, tiny and dome-shaped; SSB, PSB: both absent; SR: present; SR characteristics: tiny granules mixed with larger globules and often obscuring the recurved SZ; SZ: elongate sausage-shaped structures that lie the length of sporocyst and are slightly recurved at one end (line drawing). Distinctive features of sporocyst: tiny SB, large SR, and recurved SZs.

Prevalence: In 1/7 (14%) of the type host.

Sporulation: Unknown.

Prepatent and patent periods: Unknown.

Site of infection: Unknown. Oocysts collected from feces.

Exogenous stages: Unknown.

Cross-transmission: None to date.

Pathology: Unknown.

Materials deposited: A photoholotype is deposited in the Muséum National d'Histoire Naturelle, Paris, Reference number P7378 (1–17).

Entymology: The specific name is derived from that part of Brazil in which the infected animal was captured.

Remarks: This species most closely resembles *Eimeria motelo* described from *C. denticulata*, but found in Peru (Hůrková et al., 2000). However, oocysts of *E. motelo* are more elongated than those of *E. amazonensis* (17.0×9.4 vs. 11.7×9.1) and its sporocysts are larger (8.9×4.4 vs. 6.5×3.7). Hůrková et al. (2000) suggested that the protrusions of the oocyst wall of *E. motelo* could be due to its "wrinkling," but Lainson et al. (2008) said there is "no doubt that they, and those of *E. amazonensis*, are consistent morphological features."

There are several other turtle eimerians described with oocysts that have protrusions on their walls (see Table 4). Those of *E. mitraria*, described from *Chrysemys picta marginata*, are larger (15×10 vs. 11.7×9.1), and are truncated at one end which has from 3–4 protrusions, while the other end is conical. Those of *E. stylosa* from *Trachemys scripta elegans* also are larger (16.5×13.1) and their wall has a variable number of much longer and pointed protrusions, usually two at one end and three at the other, but sometimes up to four at one end and seven at the other. Finally, oocysts of *E. jirkamoraveci* from *Batrachemys beliostemma* has oocysts similar in size to those of *E. amazonensis* (10.6×8.9), but they have three protrusions at one end and a distinct projection at the other extremity, which contrasts with the smoothly rounded end in the oocyst wall of *E. amazonensis*.

Eimeria carajasensis Lainson, Da Silva, Franco, & De Souza, 2008

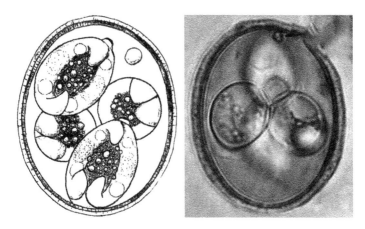

Figures 2.47, 2.48 Line drawing of the sporulated oocyst of Eimeria carajasensis *from Lainson et al. (2008), with permission from the journal* Parasite *and from the senior author. Photomicrograph of a sporulated oocyst of* E. carajasensis *from Lainson et al. (2008), with permission from the journal* Parasite *and from the senior author.*

Type host: *Chelonoidis carbonaria* (Spix, 1824) (syns. *Geochelone carbonaria* Ernst & Barbour, 1989, Red-Footed Tortoise.
Type locality: SOUTH AMERICA: (North) Brazil: Pará State, Serra dos Carajás, Salobo (6° S, 58° 18′ W).
Other hosts: Unknown.
Geographic distribution: SOUTH AMERICA: (North) Brazil: Pará State.
Description of sporulated oocyst: Oocyst shape: broadly ellipsoidal to subspheroidal; number of walls: 2; wall characteristics: brownish-yellow, with both layers striated, the inner one darker and markedly more striated than the outer one, ~1.0–1.5 thick; L × W (N = 50): 22.4 × 19.3 (20−32 × 18−25); L/W ratio: 1.2 (1.1−1.3); M, OR: both absent; PG: present, spheroidal to irregularly shaped, 2−3 wide. Distinctive features of oocyst: thick, 2-layered wall that is striated in optical cross section, lacking both M and OR, and having a distinct PG.
Description of sporocyst and sporozoites: Sporocyst shape: ovoidal; L × W (N = 50): 12 × 8 (11−13 × 7−8); L/W ratio: 1.6; SB: present, modest in size; SSB, PSB: both absent; SR: present; SR characteristics: a conspicuous structure of fine granules and larger globules not enclosed in a membrane; SZ: elongate sausage-shaped structures that lie along the length of the sporocyst and are both strongly recurved at both ends, each with two RBs. Distinctive features of

sporocyst: modest, but distinct SB, large SR, and strongly recurved SZs, each with two RBs.

Prevalence: In 2/7 (29%) of the type host; one of the infected tortoises had a concomitant infection with *E. carbonaria*.

Sporulation: Some oocysts may sporulate endogenously while others are discharged unsporulated, but complete sporulation occurs within 24 h.

Prepatent and patent periods: Unknown.

Site of infection: Unknown. Oocysts collected from feces.

Endogenous stages: Unknown.

Cross-transmission: None to date.

Pathology: Unknown.

Materials deposited: A photoholotype is deposited in the Muséum National d'Histoire Naturelle, Paris, Reference number P7378 (1−17).

Entymology: The specific name is derived from the hills of Carajás where the infected tortoises were found.

Remarks: The sporulated oocysts of *Eimeria geochelona* and *Eimeria paynei* are of similar mean size to those of *E. carajasensis*, but the former can easily be differentiated by their smooth, colorless and unstriated wall, and the latter appears to be consistently ellipsoidal in shape and has much smaller size range (19−26 × 16−20 vs. 20 × 18−32); in addition, oocysts of *E. paynei* may contain 1−3 PG and a number of smaller granules vs. the single PG of this species. Also, the SR of *E. paynei* is enclosed in a membrane, while the SR of *E. carajasensis* is not.

Eimeria carbonaria Lainson, Da Silva, Franco, & De Souza, 2008

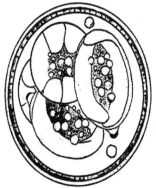 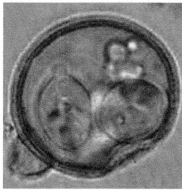

Figures 2.49, 2.50 Line drawing of the sporulated oocyst of Eimeria carbonaria *from Lainson et al. (2008), with permission from the journal* Parasite *and from the senior author. Photomicrograph of a sporulated oocyst of* E. carbonaria *from Lainson et al. (2008), with permission from the journal* Parasite *and from the senior author.*

Type host: *Chelonoidis carbonaria* (Spix, 1824) (syns. *Geochelone carbonaria* Ernst & Barbour, 1989, Red-Footed Tortoise.

Type locality: SOUTH AMERICA: (North) Brazil: Pará State, Serra dos Carajás, Salobo (6° S, 58° 18′ W).

Other hosts: Unknown.

Geographic distribution: SOUTH AMERICA: (North) Brazil: Pará State.

Description of sporulated oocyst: Oocyst shape: spheroidal to subspheroidal; number of walls: 2; wall characteristics: outer is colorless with only faint striations (pores), the inner one is brownish-yellow with conspicuous striations, ~1.0 thick; L × W (N = 50): 18.7 × 18.1 (16−21.5 × 16−21.5); L/W ratio: 1.0; M, OR: both absent; PG: present, usually one as a small spheroid and rarely two. Distinctive features of oocyst: 2-layered wall that is striated in optical cross section, lacking both M and OR, and having 1−2 distinct PGs.

Description of sporocyst and sporozoites: Sporocyst shape: ovoidal; L × W (N = 40): 11 × 7 (10−12 × 6−7); L/W ratio: 1.6; SB: present, nipple-like; SSB, PSB: both absent; SR: present; SR characteristics: a conspicuous loose mass of fine granules and larger globules; SZ: elongate sausage-shaped structures that lie along the length of the sporocyst and are moderately recurved at their ends, but without visible RBs. Distinctive features of sporocyst: distinct SB, large SR of various-sized granules and globules, and slightly recurved SZs without RBs.

Prevalence: In 2/7 (29%) of the type host; one of the infected tortoises had a concomitant infection with *E. carajasensis* and a second infected tortoise had a concomitant infection with *E. wellcomei*.

Sporulation: Exogenous. Oocysts sporulated after 3 days in 2% aqueous (w/v) potassium dichromate ($K_2Cr_2O_7$) solution at 24−26°C.

Prepatent and patent periods: Unknown.

Site of infection: Unknown. Oocysts collected from feces.

Endogenous stages: Unknown.

Cross-transmission: None to date.

Pathology: No outward signs were observed in the infected animal.

Materials deposited: A photoholotype is deposited in the Muséum National d'Histoire Naturelle, Paris, Reference number P7378 (1−17).

Entymology: The specific epithet is derived from the scientific name of the type host.

Remarks: Oocysts of this species can be distinguished from others with spheroidal to subspheroidal forms, especially *E. jaboti* and *E. lainsoni*, by their 2-layered, brownish, and striated oocyst walls.

Eimeria geochelona Couch, Stone, Duszynski, Snell, & Snell, 1996

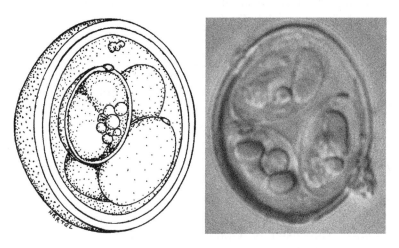

Figures 2.51, 2.52 Line drawing of the sporulated oocyst of Eimeria geochelona *from Couch et al. (1996), with permission from the* Journal of Parasitology *and from the senior author. Photomicrograph of a sporulated oocyst of* E. geochelona *from Couch et al. (1996), with permission from the* Journal of Parasitology *and from the senior author.*

Type host: *Chelonoidis nigra* (Quoy & Gaimard, 1824) (syn. *Geochelone nigra* [Quoy and Gaimard, 1824]), Charles Island Giant Tortoise.

Type locality: SOUTH AMERICA: Ecuador: Galápagos Archipelago, Isla Santa Cruz, 0°37'0" S, 90°21'0" W.

Other hosts: Unknown.

Geographic distribution: SOUTH AMERICA: Ecuador: Galápagos Archipelago, Isla Santa Cruz.

Description of sporulated oocyst: Oocyst shape: ellipsoidal to ovoidal; number of walls: 2; wall characteristics: ~ 1.5 thick, composed of 2 smooth, clear layers; outer $\sim \frac{1}{2}$ the size of inner wall; L × W (N = 50): 21.6 × 18.1 (18−25 × 16−20); L/W ratio: 1.2 (1.1−1.4); M, OR: both absent; PG: present, as 1−2 distinct large, irregularly shaped refractile bodies. Distinctive features of oocyst: presence of large, refractile PGs.

Description of sporocyst and sporozoites: Sporocyst shape: ellipsoidal; L × W (N = 50): 10.7 × 7.0 (8−12 × 5−8); L/W ratio: 1.5

(1.3×1.9); SB: present as a small nipple-like structure; SSB, PSB: both absent; SR: present; SR characteristics: medium to large granules randomly distributed among SZ; SZ: lie side-by-side along length of sporocyst. Distinctive features of sporocyst: tiny SB and large SR.

Prevalence: In 1/26 (4%) of *G. nigra* from the Charles Darwin Research Station on Isla Santa Cruz.

Sporulation: Unknown. Oocysts were kept in $K_2Cr_2O_7$ solution in the field for 3 weeks before returning them to the lab.

Prepatent and patent periods: Unknown.

Site of infection: Unknown. Oocysts collected from feces.

Exogenous stages: Unknown.

Cross-transmission: None to date.

Pathology: Unknown.

Materials deposited: Phototype of sporulated oocysts in the US National Parasite Collection (USNPC) No. 85499.

Entymology: The specific epithet is derived from the specific generic name of the definitive host (formerly *Geochelone*).

Remarks: McAllister & Upton (1989) summarized the coccidian species described from all turtles up to 1989. Sporulated oocysts of *Eimeria geochelona* resemble in shape and size those of *E. paynei* from the gopher tortoise, *Gopherus polyphemus*, from Georgia, USA and of *E. lainsoni* (syn. *E. carinii*) from the "jaboti" tortoise *Gopherus denticulata* from Brazil (see Hůrková et al., 2000, for emending *E. carinii* to *E. lainsoni*, a suggestion with which we agree). They differ from *E. paynei* in having somewhat smaller sporocysts (10.7×7.0 vs. 13.2×8.1) and in having randomly distributed, medium to large granules for their SR, rather than a mass of small granules enclosed by a thin membrane. They differ from *E. lainsoni* (syn. *E. carinii*) by having slightly larger oocysts (21.6×18.1 vs. 19.2×18.6), by lacking a distinct OR, by the presence of a PG, and by having slightly larger sporocysts (10.7×7.0 vs. 8.8×7.3) that contain an SB, which those of *E. lainsoni* lack.

Couch et al. (1996) noted that ordinarily it is not desirable to describe a new species when it is found only in one host animal because there is a likelihood the coccidian may be a pseudoparasite. However, they argued that they described this species for the following reasons: (1) there were large numbers of oocysts being discharged by the tortoise and all of the oocysts studied had good structural integrity, which suggested to them that it was a true infection; (2) the tortoises sampled were kept in pens, where contact with most other animals (that may be passing oocysts to result in spurious infections) was limited; (3) only vegetation contaminated by bird or possibly rodent feces might be a source for extraneous oocysts; however, passerine birds have mostly isosporan infections and rodents are not endemic to Isla Santa Cruz; (4) this coccidium did not resemble the eimerian oocysts from introduced rodent pest species, e.g., *Rattus* spp.; and (5) these tortoises are strict vegetarians (H.L. Snell & P.A. Stone, pers. obs.) and are not inclined to practice copraphagy.

Eimeria iversoni McAllister, Duszynski, & Roberts, 2014

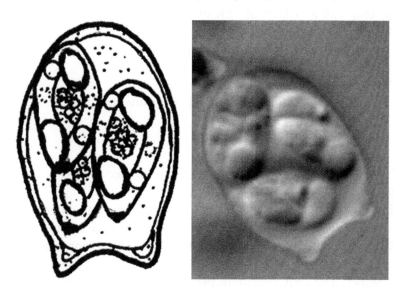

Figures 2.53, 2.54 Line drawing of the sporulated oocyst of Eimeria iversoni *from McAllister et al. (2014), with permission from the* Journal of Parasitology *and from the senior author. Photomicrograph of a sporulated oocyst of* E. iversoni *from McAllister et al. (2014), with permission from the* Journal of Parasitology *and from the senior author.*

Type host: *Chelonoidis* sp. Fitzinger, 1835, Galápagos Tortoise.
Type locality: NORTH AMERICA: USA: Texas, Dallas County, Dallas Zoo (32.74105° N, 96.817631° W).
Other hosts: None to date.

Geographic distribution: NORTH AMERICA: USA: Texas.

Description of sporulated oocyst: Oocyst shape: ovoidal; number of walls: 1; wall characteristics: smooth, ~0.5–0.8 thick; L × W: 13.5 × 10.3 (12–15 × 9–11), L/W: 1.3 (1.2–1.4); M, OR, PG: all absent. Distinctive feature of oocyst: two conical projections present on one end of oocyst, 1.0–1.5 long.

Description of sporocyst and sporozoites: Sporocyst shape: elongate-ellipsoidal, ~0.5 thick, with a smooth, single-layered wall; L × W: 8.3 × 4.5 (8–9 × 4–5), L/W ratio: 1.8 (1.7–2.2); SB: present, ~0.5 high; SSB, PSB: both absent; SR: present as 2–5 small globules; SZ: banana-shaped, 9.5 × 2.5 (9–11 × 2–3) *in situ*, with spheroidal anterior RB, 1.3 (1.0–1.6), and an ellipsoidal posterior RB, 3.2 × 2.3 (3–4 × 2.0–2.4); the N is visible, slightly posterior to midpoint of SZ. Distinctive features of sporocyst: slightly pointed at one end with a distinct SB, SR of only a few globules, and two very distinct RBs in each SZ.

Prevalence: In 1/1 (100%) of the type host.

Sporulation: Unknown.

Prepatent and patent periods: Unknown.

Site of infection: Unknown. Oocysts recovered from feces.

Exogenous stages: Unknown.

Cross-transmission: None to date.

Pathology: Unknown.

Materials deposited: Photosyntype of sporulated oocysts are deposited as USNPC, No. 106943.

Entymology: The specific epithet is a patronym to honor of Dr. John B. Iverson, Professor of Biology (retired), Earlham College, Richmond, Indiana, in recognition of his many contributions to turtle biology.

Remarks: The only coccidian previously reported from Galápagos tortoises was *Eimeria geochelona* (Couch et al., 1996). Their eimerian has ellipsoidal to ovoidal oocysts that are completely dissimilar to *E. iversoni*. However, six eimerians from turtles have been described that are more similar to *E. iversoni* because they also possess conical projections (termed mitra-like appearance) on at least one end of their oocysts and/or are irregularly shaped. These are: (1) *Eimeria amazonensis* Lainson, Da Silva, Franco, & De Souza, 2008, from the red-footed tortoise, *Chelonoidis carbonaria* from Brazil (Lainson et al., 2008); (2) *Eimeria hynek-prokopi* (Široký & Modrý, 2010), from the Indochinese box turtle, *Cuora galbinifrons* from Vietnam (Široký & Modrý, 2010); (3) *Eimeria jirkamoraveci* (Široký, Kamler, & Modrý, 2006), from Amazon toadhead turtle, *Mesoclemmys* (=*Batrachemys*) *helios-temma* from Peru (Široky et al., 2006); (4) *Eimeria mitraria* (Laveran & Mesnil, 1902) (Doflein, 1909), from numerous turtle species (>12) on three continents (McAllister & Upton, 1989; Lainson & Naiff, 1998; Segade et al., 2004; Široky et al., 2006); (5) *Eimeria motelo* (Hůrková, Modrý, Koudela, & Šlapeta, 2000) from yellow-footed tortoise *Chelonoidis* (=*Geochelone*) *denticulata* from Peru (Hůrková et al., 2000); and (6) *Eimeria stylosa* (McAllister & Upton, 1989) from red eared sliders, *Trachemys scripta elegans* from Texas, USA (McAllister & Upton, 1989). Sporulated oocysts of *E. amazonensis* have slightly smaller oocysts and sporocysts than *E. iversoni* and sporocysts of the latter have a slightly larger L/W ratio (1.8 vs. 1.7). The SR and RBs of the SZs have the most obvious differences between these two species. In *E. iversoni*, the SR has only 2−5 small globules, while the SR in *E. amazonensis* fills about half of the sporocyst and is composed of numerous granules interspersed with larger globules; in *E. iver-soni*, two RBs are visible and the posterior RB is the most promi-nent part of the SZ, while in *E. amazonensis*, Lainson et al. (2008) were unable to visualize any RBs. Sporulated oocysts of *E. hynekprokopi* differ from those of *E. iversoni* by being irregu-larly shaped, with a larger L/W ratio (1.8 vs.1.3), and having one point at one pole and two blunt points at the opposite pole. Sporulated oocysts of *E. jirkamoraveci* are smaller (10.6 × 8.9) with three conical projections (tubercles) on one end. Sporulated oocysts of *E. mitraria* differ by possessing a single conical projec-tion at one end vs. two conical projections in *E. iversoni*. Those of *E. motelo* are different in shape than those of *E. iversoni* (ellipsoi-dal vs. ovoidal), they are larger (17 × 9, L/W 1.9 vs.13.5 × 10.3, L/W 1.3) and they have small, lobed protrusions and irregularities at the poles that those of *E. iversoni* lack. Finally, sporulated oocysts of *E. stylosa* are larger (16.5 × 13.1 vs. 13.5 × 10.3) and have longer, sharper conical projections on both ends with two at one pole and three at the opposite pole.

Eimeria jaboti (Carini, 1942)

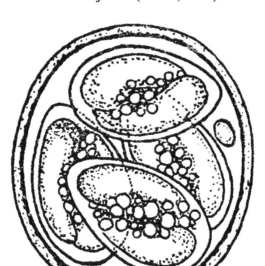

Figure 2.55 Line drawing of the sporulated oocyst of Eimeria jaboti *from Carini (1942) (journal in which it was published went bankrupt and no longer exists).*

Type host: *Chelonoidis denticulata* (L., 1766) (syns. *Geochelone denticulata* Williams, 1960; *Testudo denticulata* L., 1766; *Testudo tabulata* Duméril & Bibron, 1835, Yellow-Footed Tortoise, South American Tortoise).

Type locality: SOUTH AMERICA: (South) Brazil: São Paulo State.

Other hosts: None to date.

Geographic distribution: SOUTH AMERICA: (South) Brazil: São Paulo State.

Description of sporulated oocyst: Oocyst shape: generally spheroidal; number of walls: 3; wall characteristics: colorless, although the outer one is sometimes rough in texture (not shown in original line drawing); L × W: 17.0 wide, but some were 17–19 × 15–17; L/W ratio: 1.0, but probably ~1.1; M, OR: both absent; PG: present, conspicuous, 2–4 wide. Distinctive features of oocyst: small (mostly) spheroidal size with a distinct PG.

Description of sporocyst and sporozoites: Sporocyst shape: ovoidal; L × W: mean not given (10–11 × 6–7); L/W ratio: 1.5–1.7; SB: Carini (1942) said he did not observe one, but he did say the more tapered end "seems to be a very slight thickening of the [sporocyst] membrane"; looking at the line drawing, we think that a small SB is present; SSB, PSB: both absent; SR: present; SR characteristics: described by Carini (1942) as some "refringent granules." SZ: blunt, sausage-shaped forms, slightly pointed at one end, without RBs (line drawing).

Prevalence: Unknown.

Sporulation: 8–10 days.

Prepatent and patent periods: Unknown.

Site of infection: Unknown.

Endogenous stages: Unknown.

Cross-transmission: Carini (1942) said that attempts to infect other *Testudo* spp. adults by making them eat feces containing mature oocysts were negative.

Pathology: Unknown.

Materials deposited: None.

Remarks: Carini (1942) described the first eimerian from this host genus (which he called *T. tabulata*) and the second ever at that time from the Testudinidae. To date, excluding *E. jaboti*, there are 10 other *Eimeria* spp. described from five turtle species in three genera of this host family that include *Chelonoidis carbonaria* (4 spp.), *C. nigra* (1 sp.), *C. denticulata* (2 spp.), and *C. sp.* (1 sp.); *Gopherus polyphemus* (1 sp.); *Testudo graeca* (1 sp.). Oocysts of *E. jaboti* have a suite of mensural and quantitative characters that resemble none of these (see Table 1, McAllister et al., 2013).

Eimeria lainsoni (Lainson et al., 1990) Hůrková, Modrý, Koudela, & Šlapeta, 2000

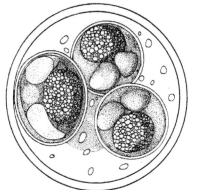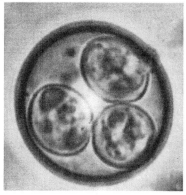

Figures 2.56, 2.57 *Line drawing of the sporulated oocyst of* Eimeria lainsoni *from Lainson et al. (1990), with permission from the* Memorias do Instituto Oswaldo Cruz *and from the senior author. Photomicrograph of a sporulated oocyst of* E. lainsoni *from Lainson et al. (1990), with permission from the* Memorias do Instituto Oswaldo Cruz *and from the senior author.*

Synonym: *Eimeria carinii* Lainson, Costa, & Shaw, 1990.
Type host: *Chelonoidis denticulata* (L., 1766) (syns. *Geochelone denticulata*, Williams, 1960; *Testudo denticulata* L., 1766), Yellow-Footed Tortoise or "jaboti" tortoise.
Type locality: SOUTH AMERICA: (North) Brazil: Pará, Serra dos Carajás (primary forest).
Other hosts: None to date.
Geographic distribution: SOUTH AMERICA: (North) Brazil: Pará.
Description of sporulated oocyst: Oocyst shape: spheroidal to subspheroidal; number of walls: 1; wall characteristics: smooth, colorless, ~1.2 thick; L × W (N = 50): 19.2 × 18.6 (15−20 × 14−19); L/W ratio: 1.0 (1.0−1.1); M, PG: both absent; OR: present, composed of a variable number, usually 10−20, of small, scattered granules, each up to ~1.0 wide, and "in constant Brownian movement." Distinctive features of oocyst: 10−20 scattered granules of the OR that appear to be in constant movement.
Description of sporocyst and sporozoites: Sporocyst shape: broadly ellipsoidal; L × W (N = 50): 8.8 × 7.3 (8−9 × 7−7.5); L/W ratio: 1.2 (1.1−1.3); SB, SSB, PSB: all apparently absent; SR: present; SR characteristics: a bulky, spheroidal to ellipsoidal mass of granules lying between the two SZ; SZ: recurved at their ends with each having both an anterior and a posterior RB and a centrally placed N. Distinctive features of sporocyst: massive SR composed of many granules.
Prevalence: Unknown in the wild; in 2/5 (40%) hosts that had been maintained together within the same compound "for some time" (Lainson et al., 1990).
Sporulation: Exogenous. Fresh feces with oocysts were lightly triturated in 2% (w/v) aqueous potassium dichromate ($K_2Cr_2O_7$) solution in covered Petri dishes held at ~24−26°C, and oocysts were observed to sporulate in 48 h.
Prepatent and patent periods: Unknown.
Site of infection: Uncertain. No oocysts were found in the gall bladder contents.
Endogenous stages: Unknown.
Cross-transmission: None to date.
Pathology: Unknown.
Materials deposited: Oocysts preserved in 10% formol-saline are being held in the Department of Parasitology, Instituto Evandro Chagas, Belém, Pará, Brazil.

Entymology: This species originally was named in honor of the late Professor A. Carini by Lainson et al. (1990), but see *Remarks*, below.

Remarks: Several eimerians described from chelonids have similarities to this species that include spheroidal to subspheroidal oocysts with an OR, but no PG or M. Oocysts of this species differ from those of *E. dericksoni* by having larger oocysts (19.2 × 18.6 vs. 10.8 × 10), have sporocysts with an SB, which those of *E. lainsoni* (syn. *E. carinii*) lack, and by being found in different genera in different families. They differ from those of *E. trionyxae* by having larger subspheroidal oocysts (19.2 × 18.6) vs. spheroidal ones (16.5 [14−18]), by their broadly ellipsoidal sporocysts vs. those which are longer and piriform (8.8 × 7.3 [8−9 × 7−7.5], L/W 1.2 vs. 12.4 × 6.2, L/W 2.0), and by being found in different genera within different families. Their oocysts differ from those of *E. lutotestudinis* by having larger oocysts (vs. 11.9 × 10.8), having a dispersed OR of many granules vs. an OR which is a membrane-bound mass of granules in *E. lutotestudinis*, and by being found in different genera within different families. Finally, oocysts described by Carini (1942) as *E. jaboti* from this same host species in São Paulo State (South) Brazil were described as spheroidal to subspheroidal, 17.0 (17−19 × 15−17), with an oocyst wall composed of three layers, the outer one sometimes with a rough texture (but not shown in Carini's line drawing). Other differences between these two species include the presence of a distinct PG (2−4 wide) vs. the multiple bodies of the OR in *E. lainsoni* (syn. *E. carinii*), the ovoidal sporocysts of *E. jaboti* (10−11 × 6−7, L/W 1.5−1.7) vs. the broadly ellipsoidal sporocysts of *E. lainsoni* (syn. *E. carinii*) (8.8 × 7.3, L/W 1.2), the vast difference in sporulation time (2 d vs. 8−10 days), and by being found in different genera within different families.

Lainson et al. (1990) described *E. carinii* from the yellow-footed tortoise, *C.* (syn. *G.*) *denticulata* from Brazil, but this name was preoccupied by a homonym, *E. carinii* Pinto, 1928, which was given to an eimerian from the rat, *Rattus norvegicus*. This prompted Hůrková et al. (2000) to rename the form named by Lainson et al. (1990) as *E. lainsoni nom. nov.*, to honor Prof. Ralph Lainson, who has done an enormous amount of work on coccidian (and other) parasites found in Brazilian vertebrates. This is a suggestion which we happily endorse.

Eimeria motelo Hůrková, Modrý, Koudela, & Šlapeta, 2000

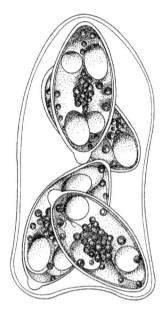

Figures 2.58, 2.59 Line drawing of the sporulated oocyst of Eimeria motelo *from Hůrková et al. (2000), with permission from the* Memorias do Instituto Oswaldo Cruz *and from D. Modrý. Photomicrograph of a sporulated oocyst of* E. motelo *from Hůrková et al. (2000), with permission from the* Memorias do Instituto Oswaldo Cruz *and from D. Modrý.*

Type host: *Chelonoidis denticulata* (L., 1766) (syns. *Geochelone denticulata* Williams, 1960; *Testudo denticulata* L., 1766), Yellow-Footed Tortoise.

Type locality: SOUTH AMERICA: Peru: Departamento de Loreto, Tamshiyacu, Iquitos (3° 59 S, 73° 10′ W).

Other hosts: Unknown.

Geographic distribution: SOUTH AMERICA: Peru: Departamento de Loreto.

Description of sporulated oocyst: Oocyst shape: irregularly ellipsoidal with slightly expressed lobed protrusions and irregularities at the poles, possibly caused by wrinkling of the wall; number of walls: 1; wall characteristics: ~0.5 thick, smooth; L × W: 17.0 × 9.4 (15–19 × 8.5–11); L/W ratio: 1.8 (1.5–2.0); M, OR, PG: all absent. Distinctive features of oocyst: the irregular shape with polar protrusions and lack of M, OR, and PG.

Description of sporocyst and sporozoites: Sporocyst shape: ellipsoidal; L × W: 8.9 × 4.4 (7.5−10 × 4−5); L/W ratio: 2.0 (1.7 × 2.5); SB: present, dome-like; SSB: described as present and indistinct, but not shown in the original line drawing; PSB: absent; SR: present; SR characteristics: small granules of irregular size, organized either in a globular cluster or distributed among SZ, or both; SZ: lie side-by-side along length of sporocyst, each with two RBs at the ends of the SZ (line drawing). Distinctive features of sporocyst: large, dome-like SB and prominent SR of many large globules.

Prevalence: In 2/4 (50%) in the type host.

Sporulation: Probably endogenous because sporulated oocysts were seen as early as 3−4 h after defecation.

Prepatent and patent periods: Unknown.

Site of infection: Unknown. Oocysts collected from feces.

Exogenous stages: Unknown.

Cross-transmission: None to date.

Pathology: Unknown.

Materials deposited: Phototypes are deposited at the Institute of Parasitology, Academy of Sciences of the Czech Republic, Branišovská 31, České Budějovice, Czech Republic (No R 111/99).

Entymology: The specific epithet *motelo* reflects the local name of the host "motelo" and is given, in accordance with the International Code of Zoological Nomenclature (Article 31.1), as a noun in apposition (ICZN, 1999).

Remarks: The yellow-footed tortoise in known locally in northwestern South America as "jaboti," "motelo," "morrocoy," or "sigrilpatoe;" it is a middle-sized, herbivorous tortoise that lives in dense rain forests, but they are often kept in captivity as pets. Lainson et al. (2008) also recorded this parasite from *C. denticulata* in Brazil. The oocysts and sporocysts they studied agreed closely with those in the original description by Hůrková et al. (2000), 17.1 × 10.8 vs. 17 × 9.4 and 9 × 5 vs. 8.9 × 4.4, respectively.

Oocysts of this species most closely resemble those of *E. amazonensis* described from *C. denticulata*, but in Brazil (Lainson et al., 2008). However, oocysts of *E. motelo* are more elongated than those of *E. amazonensis* (17.0 × 9.4 vs. 11.7 × 9.1) and its sporocysts are larger

(8.9 × 4.4 vs. 6.5 × 3.7). Hůrková et al. (2000) suggested that the protrusions of the oocyst wall of *E. motelo* could be due to its "wrinkling," but Lainson et al. (2008) said there is "no doubt that they, and
those of *E. amazonensis*, are consistant morphological features."

Eimeria welcomei Lainson, Da Silva, Franco, & De Souza, 2008

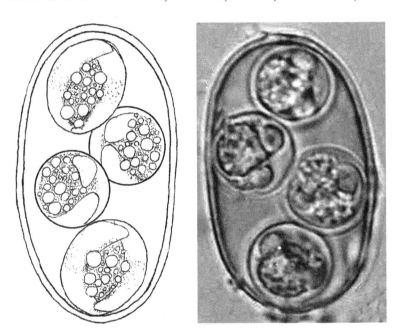

Figures 2.60, 2.61 Line drawing of the sporulated oocyst of Eimeria welcomei *from Lainson et al. (2008), with permission from the journal* Parasite *and from the senior author. Photomicrograph of a sporulated oocyst of* E. welcomei *from Lainson et al. (2008), with permission from the journal* Parasite *and from the senior author.*

Type host: *Chelonoidis carbonaria* (Spix, 1824) (syn. *Geochelone car-
bonaria* Ernst & Barbour, 1989), Red-Footed Tortoise.
Type locality: SOUTH AMERICA: (North) Brazil: Pará State,
Serra dos Carajás, Salobo (6° S, 58° 18′ W).
Other hosts: Unknown.
Geographic distribution: SOUTH AMERICA: (North) Brazil: Pará
State.
Description of sporulated oocyst: Oocyst shape: ellipsoidal to
cylindroidal; number of walls: 1; wall characteristics: colorless,
composed of a single, nonstriated layer, ~1.0 thick; L × W
(N = 50): 30.3 × 16.4 (28–32 × 15–17); L/W ratio: 1.8 (1.8–2.0);

M, OR, PG: all absent. Distinctive features of oocyst: distinctly cylindroidal shape, with a large L/W ratio and lacking M, OR, and PG.

Description of sporocyst and sporozoites: Sporocyst shape: subspheroidal to broadly ellipsoidal; L × W (N = 50): 9.6 × 7.9 (9−10 × 7−9); L/W ratio: 1.2 (1.1−1.3); SB, SSB, PSB: all absent; SR: present; SR characteristics: a mass of fine granules and larger globules in center of sporocyst; SZ: elongate sausage-shaped structures that lie the length of sporocyst and are slightly recurved at their ends, but without visible RBs. Distinctive features of sporocyst: lack of SB, large SR of various-sized granules and globules and slightly recurved SZs without RBs.

Prevalence: In 1/7 (14%) of the type host; the infected tortoises had a concomitant infection with *E. carbonaria*.

Sporulation: Unknown. Feces collected from this host could not be examined until 1 week after collection, when the oocysts were fully sporulated.

Prepatent and patent periods: Unknown.

Site of infection: Unknown. Oocysts collected from feces.

Endogenous stages: Unknown.

Cross-transmission: None to date.

Pathology: No outward signs were observed in the infected animal.

Materials deposited: A photoholotype is deposited in the Muséum National d'Histoire Naturelle, Paris, Reference number P7378 (1−17).

Entymology: The specific epithet is derived from that of Sir Henry Wellcome and in gratitude to the Wellcome Trust, London, which has a lengthy funding history for Dr. Lainson's research in Brazil.

Remarks: Lainson et al. (2008) noted that "Ellipsoidal to cylindrical oocysts, and sporocysts which lack an SB, are features of the genus *Choleoeimeria* in the biliary epithelium of reptiles (Paperna & Landsberg, 1989). *E. welcomei*...is, therefore, a likely candidate for future transference to that genus." We think this is important for future investigators to remember and to look for. There are at least a few other named eimerians from turtles which have elongate-ellipsoidal oocysts and their sporocysts either seem to lack SBs or they were described as tiny, perhaps even imagined; these may include *E. texana*, *E. welcomei*, *E. brodeni*, *E. amydae*, and *E. peltocephali* (Chapter 3).

Isospora rodriguesae Lainson, Da Silva, Franco, & De Souza, 2008

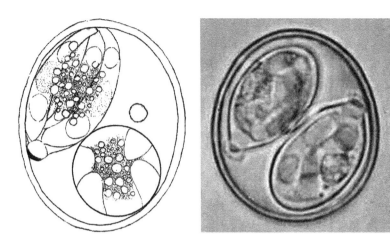

Figures 2.62, 2.63 Line drawing of the sporulated oocyst of Isospora rodriguesae *from Lainson et al. (2008), with permission from the journal* Parasite *and from the senior author. Photomicrograph of a sporulated oocyst of* I. rodriguesae *from Lainson et al. (2008), with permission from the journal* Parasite *and from the senior author.*

Type host: *Chelonoidis denticulata* (L., 1766) (syns. *Geochelone denticulata* Williams, 1960; *Testudo denticulata* L., 1766), Yellow-Footed Tortoise.

Type locality: SOUTH AMERICA: (North) Brazil: Pará State, Serra dos Carajás, Salobo (6° S, 58° 18′ W).

Other hosts: Unknown.

Geographic distribution: SOUTH AMERICA: (North) Brazil: Pará State.

Description of sporulated oocyst: Oocyst shape: broadly ellipsoidal to subspheroidal; number of walls: 1; wall characteristics: colorless, smooth, ~1.0 thick; L × W (N = 50): 24.5 × 21.0 (23−26 × 22); L/W ratio: 1.2; M, OR: both absent; PG: present, usually spheroidal, 2 wide, frequently appears to adhere to one of the sporocysts. Distinctive features of oocyst: smooth outer wall and having a distinct PG that seems to adhere to one of the sporocysts.

Description of sporocyst and sporozoites: Sporocyst shape: pear-shaped; L × W (N = 40): 16 × 9.5 (15−17 × 9−10); L/W ratio: 1.7; SB: present, cap-like, ~2 wide × 0.5 high; SSB: present, ~2.5 wide × 1−2 high; PSB: absent; SR: present; SR characteristics: a loose collection of fine granules and larger globules; SZ: elongate sausage-shaped structures that occupy the length of the sporocyst and are slightly recurved at their ends, each with an anterior and a posterior RB. Distinctive features of sporocyst: distinct SBs, large SR of various-sized granules and globules, and slightly recurved SZs, each with two RBs.

Prevalence: In 1/8 (12.5%) of the type host.

Sporulation: Exogenous. Oocysts sporulated after 24 h in 2% aqueous (w/v) potassium dichromate ($K_2Cr_2O_7$) solution at 24–26°C.

Prepatent and patent periods: Unknown.

Site of infection: Unknown. Oocysts collected from feces.

Endogenous stages: Unknown.

Cross-transmission: None to date.

Pathology: No outward signs were observed in the infected animal.

Materials deposited: A photoholotype is deposited in the Muséum National d'Histoire Naturelle, Paris, Reference number P7378 (1–17).

Entymology: The specific name of this parasite was derived as a token of thanks to Dr. Izabel R. de C. Rodrigues, who captured the type host.

Remarks: There are only two other recorded isosporans in turtles: one is *I. testudae* described from *T. horsfieldi* in Ubekistan, and the other is *I. chelydrae* from the common snapping turtle, *Chelydra serpentina*, from Arkansas in North America. The former has spheroidal oocysts with a bilayered wall, lacks a PG, and its sporocysts are devoid of an SB, all of which clearly differentiate it from the oocysts of this species (Lainson et al., 2008). Oocysts of the latter have an asymmetrical thin wall (~0.2–0.4 thick) and only measure 9.6 × 6.8 (9–11 × 6–8), lack M, OR, and PG, and have one conical projection on one side of the oocyst and two conical projections on the opposite side of the oocyst, each 1.5–1.6 long. These characters certainly distinguish it from the oocysts of *I. rodriguesae* (McAllister et al., 1990b).

Genus *Chersina* Schweigger, 1812 (Monospecific)

To our knowledge, there is a single intranuclear "coccidian" described from one *Chersina angulata* (Schweigger, 1812), South African Bowsprit or Angulate Tortoise from a zoo in Central Florida, but it has not yet been identified to genus (Garner et al., 2006). See our discussion under *Species Inquirendae* (Chapter 5).

Genus *Geochelone* Fitzinger, 1835 (3 Species)

To our knowledge, there are no coccidia described from this genus.

Genus *Gopherus* Rafinesque, 1832 (5 Species)

Eimeria paynei Ernst, Fincher, & Stewart, 1971

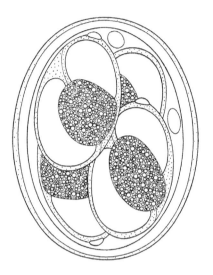

Figure 2.64 Line drawing of the sporulated oocyst of Eimeria paynei *from Ernst et al. (1971), with permission from* Comparative Parasitology *(formerly* Proceedings of the Helminthological Society of Washington*).*

Type host: *Gopherus polyphemus* (Daudin, 1808), Gopher Tortoise.
Type locality: NORTH AMERICA: USA: Georgia, Tift County.
Other hosts: None to date.
Geographic distribution: NORTH AMERICA: USA: Georgia.
Description of sporulated oocyst: Oocyst shape: ellipsoidal; number of walls: 2 (but only 1 in line drawing); wall characteristics: the outer layer lightly pitted, brownish-yellow, ~0.5 thick on the sides, thinning to ~0.25 on the ends; inner layer is colorless to light brown, ~1 thick; L × W (N = 100): 23.2 × 18.6 (19−26 × 16−20); L/W ratio: 1.25 (1.1−1.4); M, OR: both absent; PG: 1−3 are present as ellipsoidal or subspherical granules along with other granules, probably PG fragments. Distinctive features of oocyst: 1−3 PGs plus multiple other fragments.
Description of sporocyst and sporozoites: Sporocyst shape: ovoidal; L × W (N = 100): 13.2 × 8.1 (12−14 × 7−9); L/W ratio: 1.6; SB: present at the pointed end as a small, nipple-like structure (line drawing); SSB, PSB: both absent; SR: present; SR characteristics: spheroidal or ellipsoidal, composed of many small homogeneous granules enclosed by a thin membrane and takes up about ⅓ the volume of the sporocyst; SZs: fat (line drawing), lying lengthwise in

the sporocysts, partially curled around each other with a massive RB filling ½ or more of the volume of the SZ. Distinctive features of sporocyst: large RB in each SZ and a large SR that is membrane-bound.

Prevalence: In 1/1 (100%) of the type host.

Sporulation: Unknown. Oocysts were already sporulated when first examined 3 weeks after being left in 2.5% aqueous (w/v) potassium dichromate ($K_2Cr_2O_7$) solution, presumably at room temperature.

Prepatent and patent periods: Unknown.

Site of infection: Unknown.

Endogenous stages: Unknown.

Cross-transmission: None to date.

Pathology: None observed in the infected host.

Materials deposited: None.

Entymology: *Eimeria paynei* was named in honor of Dr. Jerry A. Payne, USDA Southeastern Fruit and Nut Tree Research Station, Byron, Georgia. Dr. Payne collected many of the turtles used in their parasite survey (Ernst et al., 1971).

Remarks: Oocysts of this species resemble only *E. brodeni* from the Greek tortoise, *Testudo graeca*, and those of *E. jaboti* known from *Chelonoidis denticulate*, the South American or yellow-footed tortoise (Carini, 1942). The sporulated oocysts of *E. paynei* differ from those of *E. brodeni* by having shorter oocysts (19–26 vs. 28–32), larger sporocysts (12–14 × 7–9 vs. 10 × 6–7), and by lacking a M. Ellipsoidal sporulated oocysts of *E. paynei* differ from the subspheroidal ones *E. jaboti* being larger (19–26 × 16–20 vs. 17–19 × 15–17) with obvious L/W ratio differences. In addition, the sporocysts of *E. jaboti* do not have an SB, while those of *E. paynei* do.

Genus *Homopus* Walbaum, 1782 (5 Species)

To our knowledge, there are no coccidia described from this genus.

Genus *Indotestudo* Schlegel & Müller, 1844 (3 Species)

To our knowledge, there a single intranuclear "coccidian" described from three *Indotestudo forstenii* (Schlegel & Müller, 1844), Travancore or Foresten's Tortoises, taken from an illegal shipment originating in Celebese (Sulawesi) Island, Indonesia (Garner et al., 1998, 2006). Five additional tortoises presented with both the intranuclear coccidium and *Mycoplasma* spp. from their nasal cavities (Innis et al., 2007, and an undated, unpublished report we were able to read). Unfortunately, this coccidian has not yet been identified to genus. See our discussion under *Species Inquirendae* (Chapter 5).

Genus *Kinixys* Schweigger, 1812 (8 Species)
To our knowledge, there are no coccidia described from this genus; however, see table in Bourdeau (1989) for *Cryptosopordium* sp.

Genus *Malacochersus* Siebenrock, 1903 (Monospecific)
To our knowledge, there are no coccidia described from this genus.

Genus *Manouria* Schlegel & Müller, 1844 (2 Species)
To our knowledge, there is a single intranuclear "coccidian" described from one *Manouria impressa* (Günther, 1882), Impressed Tortoise, which had been taken from an illegal shipment originating in Celebese (Sulawesi) Island, Indonesia and been deposited in the Fort Worth Zoo, Fort Worth, Texas (Garner et al., 1998, 2006). Unfortunately, this coccidian has not yet been identified to genus. See our discussion under *Species Inquirendae* (Chapter 5).

Genus *Psammobates* Linnaeus, 1758 (3 Species)
To our knowledge, there are no coccidia described from this genus.

Genus *Pyxis* Bell, 1827 (2 Species)
To our knowledge, there are no coccidia described from this genus.

Genus *Stigmochelys* Gray, 1873 (Monospecific)
To our knowledge, there a single intranuclear "coccidian" described from one *Stigmochelys pardalis* (Bell, 1828) (syn. *Geochelone pardalis* Bour, 1980), Leopard Tortoise, taken from a private facility in Baton Rouge, Louisiana (Garner et al., 1998, 2006). Unfortunately, this coccidian has not yet been identified to genus. See our discussion under *Species Inquirendae* (Chapter 5).

Genus *Testudo* Linnaeus, 1758 (5 Species)
Eimeria brodeni Cerruti, 1930

Figure 2.65 Line drawing of the sporulated oocyst of Eimeria brodeni *from Cerruti, 1930 (journal in which it was published no longer exists).*

Type host: *Testudo graeca* L., 1758, Spur-Thighed Tortoise.

Type locality: WESTERN EUROPE: Italy: Island of Sardinia (autonomous region of Italy).

Other hosts: None to date.

Geographic distribution: WESTERN EUROPE: Italy.

Description of sporulated oocyst: Oocyst shape: ovoidal; number of walls: 2 (but only 1 in original line drawing); wall characteristics: smooth; L × W: 30 × 19 (28−32 × 18−20); L/W ratio: 1.6; M: present; OR, PG: both absent (line drawing). Distinctive features of oocyst: thin, single-layered wall that thins even more at one pole (see line drawing).

Description of sporocyst and sporozoites: Sporocyst shape: ellipsoidal; L × W: mean unknown (10 × 6−7); L/W ratio: unknown; SB: present, but very tiny; SSB, PSG: both absent; SR: present; SR characteristics: some granules; SZ: sickle-shaped. Distinctive features of sporocyst: None.

Prevalence: Unknown.

Sporulation: Oocysts typically undergo sporulation in 1−2 days, but sometimes take 5−6 days (Pellérdy, 1974).

Prepatent and patent periods: Unknown.

Site of infection: Cerruti (1930) stated that he could not precisely state in which organ the endogenous stages were found, but that he thought they were "an infestation of the intestinal mucosa." His examination of the liver and gall bladder were both negative as were some sections of the small intestine.

Endogenous stages: Unknown.

Cross-transmission: None to date.

Pathology: Unknown.

Materials deposited: None.

Entymology: The specific epithet was named in honor of Prof. Broden, who was recently deceased when Cerruti (1930) wrote the species description.

Remarks: Cerruti (1930) gave just the most marginal of descriptions and this "species" probably should be relegated to a *species inquirenda*. We include it here only because he provided a line drawing, which itself is of marginal value. This morphotype has not been recognized or redescribed since its original description 80+ years ago.

Isospora testudae Davronov, 1985

Figure 2.66 Line drawing of the sporulated oocyst of Isospora testudae *(original) modified from Davronov (1985).*

Type host: *Testudo horsfieldii* Gray, 1844, Horsfield's Tortoise.

Type locality: ASIA: Uzbekistan.

Other hosts: None to date.

Geographic distribution: ASIA: Uzbekistan.

Description of sporulated oocyst: Oocyst shape: spheroidal; number of walls: 2 (only 1 in original line drawing); wall characteristics: smooth; L × W: 25.6 (22−29); L/W ratio: 1.0; M, OR, PG: all absent. Distinctive features of oocyst: simple spheroidal body lacking M, OR, and PG.

Description of sporocyst and sporozoites: Sporocyst shape: elongate-ovoidal; L × W: mean not given (15−19 × 10−15); L/W ratio: ~1.4; SB: likely present at pointed end of sporocyst, but not shown in original line drawing; SSB, PSB: both absent; SR: present; SR characteristics: a few small granules; SZ: spheroidal (original line drawing). Distinctive features of sporocyst: elongated-ellipsoidal shape containing small, spheroidal SZ (original line drawing).

Prevalence: In 17/63 (27%) of the type host.

Sporulation: Unknown.

Prepatent and patent periods: Unknown.

Site of infection: Unknown.

Endogenous stages: Unknown.

Cross-transmission: None to date.

Pathology: Unknown.

Materials deposited: None.

Remarks: Davronov (1985) gave just a marginal description, and this "species" probably should be relegated to a *species inquirenda*. We include it here only because he provided a line drawing, which itself is of marginal value. No one else has seen an oocyst resembling the one Davronov (1985) provided, and it seems likely to us that it could be a spurious finding of a bird isosporan.

FAMILY GEOEMYDIDAE (BATAGURIDAE) ASIAN RIVER, LEAF & ROOFED, & ASIAN BOX TURTLES, 19 GENERA, 70 SPECIES

Genus *Batagur* Gray, 1856 (6 Species)

Eimeria zbatagura Široky & Modrý, 2010

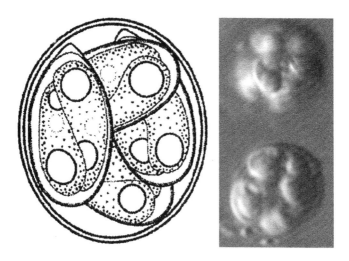

Figures 2.67, 2.68 Line drawing of the sporulated oocyst of Eimeria zbatagura *from Široky and Modrý (2010), with permission from* Acta Protozoologica *and from both authors. Photomicrograph of a sporulated oocyst of* E. zbatagura *from Široky and Modrý (2010), with permission from* Acta Protozoologica *and from both authors.*

Type host: *Batagur baska* (Gray, 1830), River Terrapin.

Type locality: ASIA: Singapore.

Other hosts: None to date.

Geographic distribution: ASIA: Singapore.

Description of sporulated oocyst: Oocyst shape: subspheroidal to slightly ellipsoidal; number of walls: 1; wall characteristics: smooth, colorless, very thin, and fragile; L × W (N = 30): 7.4 × 6.3 (6−8 × 5−7); L/W ratio: 1.2 (1.0−1.6); M, PG, OR: all absent. Distinctive features of oocyst: tiny size, lack of M, PG, and OR, and entire interior space is packed with sporocysts.

Description of sporocyst and sporozoites: Sporocyst shape: ovoidal; L × W: 5.5 × 3 (5−6 × 2.5−3); L/W ratio: 1.9 (1.7−2.0); SB: present, as a small nipple-like projection; SSB, PSB, SR: all absent; SZ: sausage-shaped (line drawing), lie head-to-tail, each with a subspheroidal RB at rounded end, ~ 1.5 wide; N: visible only occasionally, small, <1.0 wide. Distinctive features of sporocyst: lacking SSB, PSB, and SR.

Prevalence: Not studied; the authors examined mixed samples that originated from a group of five turtles.

Sporulation: Unknown. Collected fresh feces were placed in 2.5% (w/v) $K_2Cr_2O_7$ solution and "allowed to sporulate at room temperature" (°C, not stated), but these methods did not allow Široky and Modrý (2010) to determine if the oocysts were sporulated or unsporulated when first placed into 2.5% (w/v) $K_2Cr_2O_7$.

Endogenous stages: Unknown. Oocysts were collected from the feces of live animals.

Cross-transmission: None to date.

Pathology: Unknown.

Material deposited: Photosyntypes deposited in the collection of the Institute of Parasitology, Biology Centre of the Academy of Sciences of the Czech Republic, České Budějovice, Czech Republic, No. IP ProtColl P12.

Entymology: The specific epithet was given "in accordance with the International Code of Zoological Nomenclature (Article 31.1) as a noun in apposition." It means "originating from Batagur" in Czech language (Široky & Modrý, 2010).

Remarks: According to the original authors, none of the *Eimeria* species described from any chelonian species possess the unique, tiny sporulated oocysts as does this species.

Genus *Cuora* Gray, 1856 (13 Species)
Eimeria hynekprokopi Široky & Modrý, 2010

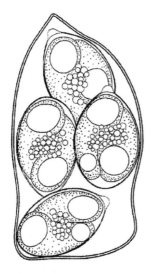 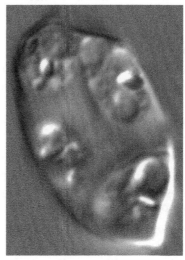

Figures 2.69, 2.70 Line drawing of the sporulated oocyst of Eimeria hynekprokopi *from Široky and Modrý (2010), with permission from* Acta Protozoologica *and from both authors. Photomicrograph of a sporulated oocyst of* E. hynekprokopi *from Široky and Modrý (2010), with permission from* Acta Protozoologica *and from both authors.*

Type host: *Cuora galbinifrons* Bourret, 1939, Indochinese Box Turtle.

Type locality: ASIA: Vietnam: central.

Other hosts: None to date.

Geographic distribution: ASIA: Vietnam.

Description of sporulated oocyst: Oocyst shape: an irregular shape that can have one or two points at both poles with the shape often influenced by the position of the sporocysts; number of walls: 1; wall characteristics: smooth, colorless, very thin, and fragile; $L \times W$ (N = 30): 15.6×8.7 $(14-18 \times 8-10)$; L/W ratio: 1.8 (1.4–2.3); M, PG, OR: all absent. Distinctive features of oocyst: amorphous shape, very thin wall, and lacking M, PG, and OR.

Description of sporocyst and sporozoites: Sporocyst shape: broadly ovoidal; $L \times W$: 6.5×4.3 $(5-8 \times 4-5)$; L/W ratio: 1.5 (1.3–1.9); SB: present, as a small nipple-like projection; SSB, PSB: both absent; SR: present; SR characteristics: numerous tiny granules scattered among SZ; SZ: sausage-shaped (line drawing), lie head-to-tail (sometimes encircling each other), each with a larger subspheroidal RB at rounded end, $\sim 2-2.5$, and a smaller one, $\sim 1.5-2$, at pointed end; N: not visible. Distinctive features of sporocyst: sausage-shaped SZ, each with one large and one small RB and SR of tiny granules.

Prevalence: In 10/34 (29%) of the type host.

Sporulation: Unknown. Collected fresh feces were placed in 2.5% (w/v) $K_2Cr_2O_7$ solution and "allowed to sporulate at room temperature" (°C, not stated), but these methods did not allow Široký and Modrý (2010) to determine if the oocysts were sporulated or unsporulated when first placed into 2.5% (w/v) $K_2Cr_2O_7$.

Endogenous stages: Unknown. Oocysts were collected from the feces of live animals.

Cross-transmission: None to date.

Pathology: Unknown.

Material deposited: Photosyntypes deposited in the collection of the Institute of Parasitology, Biology Centre of the Academy of Sciences of the Czech Republic, České Budějovice, Czech Republic, No. IP ProtColl P11.

Entymology: The specific epithet was given in honor of Hynek Prokop, for his "unselfish temporal keeping of turtles and for all the assistance with sampling during the study in which the host turtles were collected" (Široky & Modrý, 2010).

Remarks: This species belongs to the group of extremely thin-walled oocysts of turtles that include *Eimeria patta* and *E. motelo*. It differs from the former by having more elongated oocysts and sporocysts and from the latter by having longer sporocysts (L/W ratio: 1.7−2.5 vs. 1.3−1.9).

Genus *Cyclemys* Bell, 1834 (7 Species)
Eimeria palawanensis Široky & Modrý, 2010

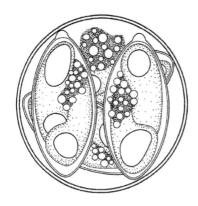 

Figures 2.71, 2.72 Line drawing of the sporulated oocyst of Eimeria palawanensis *from Široky and Modrý (2010), with permission from* Acta Protozoologica *and from both authors. Photomicrograph of a sporulated oocyst of* E. palawanensis *from Široky and Modrý (2010), with permission from* Acta Protozoologica *and from both authors.*

Type host: *Cyclemys dentata* (Gray, 1831), Asian Leaf Turtle.
Type locality: ASIA: Philippines: Palawan Island, Malinao River, ~35 km south of Puerto Princessa.
Other hosts: None to date.
Geographic distribution: ASIA: Philippines.
Description of sporulated oocyst: Oocyst shape: spheroidal; number of walls: 1; wall characteristics: colorless, smooth, fragile, very thin, ~1.0 thick; L × W: 14−17 wide; L/W ratio: 1.0−1.1; M, PG: both absent; OR: present; OR characteristics: nonmembranous, irregular-shaped body, 3 × 3−4, composed of fine granules. Distinctive features of oocyst: spheroidal shape, lacking M and PG, and large OR of fine granules.
Description of sporocyst and sporozoites: Sporocyst shape: elongate-ovoidal to ellipsoidal, slightly pointed at end opposite SB (line drawing); L × W: 12.6 × 5.8 (11−13 × 5−6); L/W ratio: 2.2 (1.8−2.4); SB: present, knob-like, 1−1.5 wide and 1 high; SSB,

PSB: both absent; SR: present; SR characteristics: several granules scattered among SZ; SZ: vermiform, lie head-to-tail, each with a larger RB, $\sim 2 \times 2-3.5$ at rounded end and a smaller RB, $1-2$ wide at pointed end; N: not visible. Distinctive features of sporocyst: slightly pointed end opposite knob-like SB.

Prevalence: In 1/6 (17%) of the type host.

Sporulation: Unknown. Collected fresh feces were placed in 2.5% (w/v) $K_2Cr_2O_7$ solution and "allowed to sporulate at room temperature" (°C, not stated), but these methods did not allow Široky and Modrý (2010) to determine if the oocysts were sporulated or unsporulated when first placed into 2.5% (w/v) $K_2Cr_2O_7$.

Endogenous stages: Unknown. Oocysts were collected from the feces of live animals.

Cross-transmission: None to date.

Pathology: Unknown.

Material deposited: Photosyntypes deposited in the collection of the Institute of Parasitology, Biology Centre of the Academy of Sciences of the Czech Republic, České Budějovice, Czech Republic, No. IP ProtColl P14. An alcohol-preserved symbiotype host is deposited in the Zoological collection of the National Museum Prague, No. NMP6V 74113.

Entymology: The specific epithet was derived from the name of Palawan Island where the type locality of this species is found.

Remarks: The fragile spheroidal oocysts of this species somewhat resemble those of *E. pseudemydis*, *E. tetradacrutata*, *E. carri*, *E. lainsoni*, and *E. ornata* from other turtle species. Oocysts of *E. pseudemydis* are slightly bigger than those of this species (19×17.5 vs. $14-17$) and are broadly ellipsoidal with pear-shaped sporocysts. Oocysts of *E. tetradacrutata* are bigger and more elongated (19.5×19.2 vs. $14-17$), have a thicker oocyst wall that is more-or-less striated, have a membrane-bound OR, and have sporocysts that are tear-drop shaped. Oocysts of *E. carri* are slightly smaller and more elongated (15.9×14.5 vs. $14-17$), have a membrane-bound OR, and have sporocysts with a tiny SB. Oocysts of *E. lainsoni* are bigger (19.2×18.6 vs. $14-17$), have a thicker oocyst wall, and have broader sporocysts that lack a visible SB. Finally, oocysts of *E. ornata* differ from this species by having more elongated oocysts, the presence of a PG, and sporocysts with a much smaller SB.

Eimeria petrasi Široký & Modrý, 2010

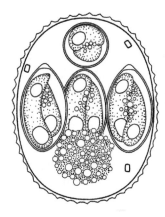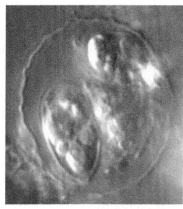

Figures 2.73, 2.74 Line drawing of the sporulated oocyst of Eimeria petrasi *from Široký and Modrý (2010), with permission from* Acta Protozoologica *and from both authors. Photomicrograph of a sporulated oocyst of* E. petrasi *from Široký and Modrý (2010), with permission from* Acta Protozoologica *and from both authors.*

Type host: *Cyclemys dentata* (Gray, 1831), Asian Leaf Turtle.

Type locality: ASIA: Philippines: Palawan Island, Malinao River, ~35 km south of Puerto Princessa.

Other hosts: None to date.

Geographic distribution: ASIA: Philippines.

Description of sporulated oocyst: Oocyst shape: broadly ellipsoidal; number of walls: 1; wall characteristics: colorless, fragile, having a wavy outer surface densely covered with small, pointed, knob-like projections; L × W (N = 20): 22.9 × 18.6 (20−25 × 16−20); L/W ratio: 1.2 (1.1−1.4); M: absent, PG: up to three present, each ~1 × 1−2; OR: present; OR characteristics: highly variable in size and structure from scattered fine granules to an irregular globular structure, ~8 wide, composed of fine granules. Distinctive features of oocyst: large size, with three PGs, and a large OR, and a sculptured outer wall.

Description of sporocyst and sporozoites: Sporocyst shape: elongate-ovoidal (line drawing), but original authors say it is pointed at both ends, a feature not seen in their line drawing or photomicrographs; L × W: 12.3 × 6.5 (11−13 × 6−7); L/W ratio: 1.9 (1.7−2.0); SB: present, as a small nipple-like projection; SSB, PSB: both absent; SR: present; SR characteristics: small granules scattered among SZ; SZ: sausage-shaped (line drawing), lie head-to-tail, each with a subspheroidal RB, ~2−3 wide, at both ends; N: not visible. Distinctive features of sporocyst: nothing; fairly typical ovoidal structure with a nipple-like SB.

Prevalence: In 1/6 (17%) of the type host.

Sporulation: Unknown. Collected fresh feces were placed in 2.5% (w/v) $K_2Cr_2O_7$ solution and "allowed to sporulate at room temperature" (°C, not stated), but these methods did not allow Široky and Modrý (2010) to determine if the oocysts were sporulated or unsporulated when first placed into 2.5% (w/v) $K_2Cr_2O_7$.

Endogenous stages: Unknown. Oocysts were collected from the feces of live animals.

Cross-transmission: None to date.

Pathology: Unknown.

Material deposited: Photosyntypes deposited in the collection of the Institute of Parasitology, Biology Centre of the Academy of Sciences of the Czech Republic, České Budějovice, Czech Republic, No. IP ProtColl P13.

Entymology: The specific epithet is in honor of Petr Petrás, for his "long-lasting and generous help with sampling and field work."

Remarks: According to the original authors, there are no eimerians from turtles that have sporulated oocysts with the wavy outer surface and multiple PGs seen in this species.

Genus *Geoclemys* Gray, 1831 (Monospecific)

To our knowledge, there are no coccidia described from this genus.

Genus *Geoemyda* Gmelin, 1789 (2 Species)

To our knowledge, there are no coccidia described from this genus.

Genus *Hardella* Gray, 1831 (Monospecific)

To our knowledge, there are no coccidia described from this genus.

Genus *Heosemys* Gray, 1831 (4 Species)

Eimeria arakanensis Široky & Modrý, 2006

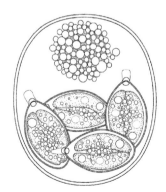 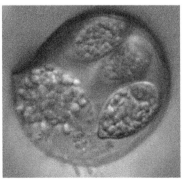

Figures 2.75, 2.76 Line drawing of the sporulated oocyst of Eimeria arakanensis *from Široky and Modrý (2006), with permission from* Acta Protozoologica *and from both authors. Photomicrograph of a sporulated oocyst of* E. arakanensis *from Široky and Modrý (2010), with permission from* Acta Protozoologica *and from both authors.*

Type host: *Heosemys depressa* (Anderson, 1875), Akaran Forest turtle.

Type locality: ASIA: Western Myanmar (Burma).

Other hosts: *Cuora flavomarginata* (Gray, 1863), Yellow-Margined Box Turtle.

Geographic distribution: ASIA: China, Western Myanmar (Burma).

Description of sporulated oocyst: Oocyst shape: broadly ovoidal to subspheroidal; number of walls: 1; wall characteristics: smooth, colorless, ~ 0.6 thick; L × W: 28.0 × 23.6 (24−30 × 22−25); L/W ratio: 1.2 (1.1−1.4); M, PG: both absent; OR: present; OR characteristics: globular, 10−15 wide, composed of fine granules. Distinctive features of oocyst: relatively large size, very thin outer wall, and presence of a large, globular OR.

Description of sporocyst and sporozoites: Sporocyst shape: elongate-ovoidal to ellipsoidal, slightly pointed at end opposite SB; L × W: 12.8 × 7.0 (12−15 × 6−8); L/W ratio: 1.9 (1.6−2.2); SB: present, knob-like, ~ 1 high × 1−2 wide and there is a very thin membranous, cap-like structure over layering it; SSB: present, a homogeneous subglobular structure, 1−1.5 high × 1.5−2 wide; PSB: absent; SR: present; SR characteristics: small granules of irregular sizes usually scattered among SZ; SZ: cucumber-shaped, each with a subspheroidal RB, 1−2 × 1.5−2, at each end; N: small, $\sim 1-2$ wide located between the two RBs. Distinctive features of sporocyst: very thin, membranous, cap-like structure over layering SB.

Prevalence: In 3/9 (33%) of type host; in 3/6 (50%) *C. flavomarginata*.

Sporulation: Exogenous. Oocysts became fully sporulated in 3−5 days at 20−23°C.

Endogenous stages: Unknown. Oocysts were collected from the feces of live animals.

Cross-transmission: None to date.

Pathology: Unknown.

Material deposited: Photosyntypes are deposited in the Department of Veterinary and Pharmaceutical Sciences, Brno, Czech Republic, No. R 100/05.

Entymology: The specific epithet is derived from the name of the Arakan hills, the "terra typical" of the type host.

Remarks: Only the sporulated oocysts of *E. brodeni*, previously described from *Testudo gracea* in Italy (Cerruti, 1930), and those of *E. chrysemydis*, from *Chrysemys picta bellii* in Iowa, USA (Deeds & Jahn, 1939), somewhat resemble those of this species. However, oocysts of *E. brodeni* are narrower and lack an OR, while those of *E. chrysemydis*, though similar in size, are pear-shaped. Široký & Modrý (2010) found this species in a second host, the yellow-margined box turtle, *C. flavomarginata*, from China; the oocysts in this species were 28.5 × 23.6 (25−31 × 21−27) with L/W ratio, 1.2 (1.1−1.4). They said the morphology of the sporulated oocysts from this second host species "fit well the traits of this species from the original description."

Genus *Leucocephalon* McCord, Iverson, & Boeadi, 1995 (Monospecific)
To our knowledge, there are no coccidia described from this genus.

Genus *Malayemys* Lindholm, 1931 (2 Species)
Eimeria surinensis Široký & Modrý, 2010

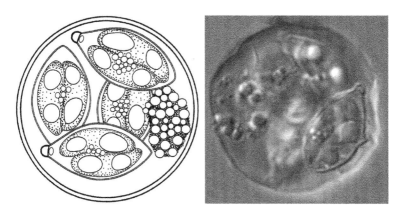

Figures 2.77, 2.78 *Line drawing of the sporulated oocyst of* Eimeria surinensis *from Široký and Modrý (2010), with permission from* Acta Protozoologica *and from both authors. Photomicrograph of a sporulated oocyst of* E. surinensis *from Široký and Modrý (2010), with permission from* Acta Protozoologica *and from both authors.*

Type host: *Malayemys subtrijuga* (Schlegel & Müller, 1845), Malayan Snail-Eating Turtle.
Type locality: ASIA: Thailand: Isaan Province, Surin.
Other hosts: None to date.
Geographic distribution: SOUTHEAST ASIA: Thailand.
Description of sporulated oocyst: Oocyst shape: spheroidal to subspheroidal; number of walls: 2; wall characteristics: smooth, colorless, ∼1.5 thick; L × W (N = 30): 22.6 × 21.4 (21−25 × 20−25); L/W ratio: 1.1 (1.0−1.2); OR: present, distinct;

OR characteristics: globular structure, 5−7 wide, composed of small globules and fine granular material; M, PG: both absent. Distinctive features of oocyst: spheroidal, with a thin wall and a large, globular OR.

Description of sporocyst and sporozoites: Sporocyst shape: spindle-shaped to elongate-ellipsoidal, but pointed at both ends; L × W: 13.4 × 6.9 (13−15 × 6−8); L/W ratio: 1.9 (1.8−2.3); SB: present, robust, cup-like, 2 × 1.5−2; SSB: present, a homogeneous sub-globular structure, 1.5−2 × 1−1.5; PSB: absent; SR: present; SR characteristics: granular material scattered among SZ; SZ: cucumber-shaped, each with a subspheroidal RB, ∼2.3 × 2−3, at each end. Distinctive features of sporocyst: narrow ellipsoid, pointed at both ends with both an SB and an SSB.

Prevalence: In 4/4 (100%) of the type host.

Sporulation: Unknown. Collected fresh feces were placed in 2.5% (w/v) $K_2Cr_2O_7$ solution and "allowed to sporulate at room temperature" (°C, not stated), but these methods did not allow Široky and Modrý (2010) to determine if the oocysts were unsporulated when first placed into 2.5% (w/v) $K_2Cr_2O_7$.

Endogenous stages: Unknown. Oocysts were collected from the feces of live animals.

Cross-transmission: None to date.

Pathology: Unknown.

Material deposited: Photosyntypes deposited in the collection of the Institute of Parasitology, Biology Centre of the Academy of Sciences of the Czech Republic, České Budějovice, Czech Republic, No. IP ProtColl P9. Symbiotype host is an alcohol-preserved voucher specimen in the Zoological collection of the National Museum Prague, collection number NMP6V 73559.

Entymology: The specific epithet was derived from Surin, the "terra typical" of this species.

Remarks: This species distantly resembles *E. tetradacrutata* and *E. carbonaria*. Its sporulated oocysts differ from those of the former by having larger oocysts (22.6 × 21.4 vs. 19.5 × 19.2), with a thinner oocyst wall (1−1.5 vs. 1.3−1.9) that is bilayered and colorless vs. single-layered and striated. Also, sporulated oocysts of *E. tetradacrutata* have a membrane-bound OR and their SBs are very different (Wacha & Christiansen, 1976). Sporulated oocysts of *E. carbonaria* are smaller than those of this species (18.7 × 18.1 vs. 22.6 × 21.4) and they lack an OR which is present in sporulated oocysts of this species (Lainson et al., 2008).

Genus *Mauremys* Gray, 1869 (9 Species)

Eimeria mitraria (Laveran & Mesnil, 1902) Doflein, 1909

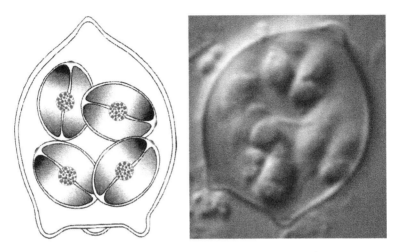

Figures 2.79, 2.80 Line drawing of the sporulated oocyst of Eimeria mitraria *from Široký and Modrý (2006), with permission from* Acta Protozoologica *and from both authors. Photomicrograph of a sporulated oocyst of* E. mitraria *from Široký and Modrý (2006), with permission from* Acta Protozoologica *and from both authors.*

Synonyms: *Coccidium mitrarium* Laveran & Mesnil, 1902; *Eimeria mitrarium* Deeds & Jahn, 1939.

Type host: *Mauremys reevesii* (Gray, 1831) (syn. *Chinemys reevesii* Smith, 1931), Reeves' Turtle.

Type locality: ASIA: Japan (?).

Other hosts: *Chelydra serpentina serpentina* (L., 1758), Common Snapping Turtle; *Chrysemys picta belli* (Gray, 1831), Western Painted Turtle; *Emydoidea blandingii* (Holbrook, 1838), Blandings Turtle; *Emys orbicularis* (L., 1758), European Pond Turtle; *Graptemys geographica* (Le Sueur, 1817), Common Map Turtle; *Graptemys pseudogeographica* (Gray, 1831), False Map Turtle; *Graptemys versa* Stejneger, 1925, Texas Map Turtle; *Heosemys depressa* (Anderson, 1875), Arakan Forest Turtle; *Kinosternon flavescens flavescens* Wermuth & Mertens, 1977, Yellow Mud Turtle; *Kinosternon flavescens spooneri* Wermuth & Mertens, 1977, Yellow (or Illinois) Mud Turtle; *Mesoclemmys heliostemma* (McCord et al., 2001), Amazon Toadhead Turtle; *Pseudemys texana* Baur, 1893, Texas River Cooter; *Terrapene carolina triunguis* (Agassiz, 1857), Three-Toed Box Turtle; *Trachemys scripta elegans* (Wied, 1838), Red Eared Slider.

Geographic distribution: ASIA: Japan (?); Myanmar; EUROPE: Spain; NORTH AMERICA: USA: Arkansas, Iowa, Texas.

Description of sporulated oocyst: Oocyst shape: truncated ovoidal with a distinct point, ~1–1.5 long, at the slightly pointed end of the ovoid while the other end is truncated into a flat base that is ornamented with three or four points, giving the oocyst a mitra-like or projectile-like appearance; number of walls: 1; wall characteristics: smooth, colorless, ~0.5 thick; L × W: 10–15 × 5–11 (Deeds & Jahn, 1939) or L × W: 10.0 × 7.6 (8–11.5 × 6–9) with L/W ratio: 1.3 (1.1–1.6) (Wacha & Christiansen, 1976); M, OR, PG: all absent. Distinctive features of oocyst: unique bullet-like shape with one spine at the rounded end and three or four spines ornamenting the flattened end.

Description of sporocyst and sporozoites: Sporocyst shape: ellipsoidal; L × W: 5.2 × 3.0 (4–6 × 3–4) with L/W ratio: 1.7 (Wacha & Christiansen, 1976); SB: present, small, conical; SSB, PSB: both absent; SR: present; SR characteristics: spheroidal, membrane-bound mass, ~2 wide, containing several spheroidal granules; SZ: head-to-tail within sporocyst each with one prominent subspheroidal RB, 1.5–2 × 2–3, at broad end. Distinctive features of sporocyst: very narrow and ellipsoid with slightly pointed ends.

Prevalence: In 2/2 (100%) of the type host; in 1/9 (11%) *T. c. triunguis* from Arkansas; in 1/3 (33%) *G. versa*, in 1/5 (20%) *K. f. flavescens*, in 2/9 (22%) *P. texana*, and in 24/100 (24%) *T. s. elegans* from Texas (McAllister et al., 1991, 1994); in 48/150 (32%) *C. p. belli* from Iowa (Deeds & Jahn, 1939), and again in 7/22 (32%) *C. p. belli* from Iowa (Wacha & Christiansen, 1976); in 1/2 (50%) *G. geographica*, 3/8 (37.5%) *G. pseudogeographica*, and in 4/26 (15%) *K. f. spooneri* all from Iowa (Wacha & Christiansen, 1976); in 9/9 (100%) *H. depressa* that were freshly caught in the wild in Myanmar, imported into China, and later imported into the Czech Republic from China (Široký & Modrý, 2006); in 11/44 (25%) *E. orbicularis* from two localities in Spain (Segade et al., 2006); unknown in two *M. heliostemma* from Peru since fecal samples from these hosts were mixed before examination (Široký et al., 2006a).

Sporulation: Likely exogenous, but really unknown. Laveran and Mesnil (1902) killed two turtles and said they found "many cysts in all states of maturation, [in] a very special form, reminiscent of a miter." Others have placed turtle feces into specimen jars in 2.5%

aqueous (w/v) $K_2Cr_2O_7$ solution at $\sim 22°C$ for 1 week and then examined. Samples with sporulated oocysts were then stored in a refrigerator at $\sim 5°C$.

Prepatent and patent periods: Unknown.

Site of infection: Small intestine (Laveran & Mesnil, 1902).

Endogenous stages: Laveran & Mesnil (1902) apparently did not fix intestinal tissue and embed it for histological examination; rather, they "followed fresh preparations." We interpret that to mean that they looked at fresh gut tissue in tissue squashes and within these preparations they saw meronts, 10–12, with 20 merozoites, each 3–5 long. Microgametocytes, 10–15 wide, contained "hair-colored" microgametes, 5–6 long that were "filled with greenish granules" (Laveran & Mesnil, 1902).

Cross-transmission: None to date.

Pathology: Unknown.

Materials deposited: Segade et al. (2006) deposited photosyntypes of sporulated oocysts in the Museo Nacional de Ciencias Naturales (CSIC), Madrid, Spain as MNCN 35.01/13.

Remarks: This species was initially found and named by Laveran and Mesnil (1902) from an Asiatic turtle native to China that was brought to their museum in Paris, France. They speculated that the turtle may have come from Japan or Ceylon (now known as Sri Lanka), but "the information provided by the merchant on the origin of the turtle was incorrect" (Laveran & Mesnil, 1902). It is curious to note that Laveran and Mesnil (1902) said that the endogenous stages they saw in the intestine were all *extracellular*. Transliteration of their statement (in French) is "Sections of the small intestine we see has all the states of growth and differentiation, more or less closely juxtaposed to the epithelial cells of which the plateau appears when they miss and depressed, often, the parasite takes an elongated shape by stretching along the epithelial surface" and "It is likely that the coccidia feed on the epithelial cell through the intermediary of pseudopodia" (Laveran & Mesnil, 1902).

This parasite was redescribed by Deeds and Jahn (1939), who found it in *C. p. belli* in Iowa. Later, *E. mitraria* was again redescribed from other *C. p. belli* from Iowa (Wacha & Christiansen, 1976) to include an SB, a structural feature previously unreported. Oocysts of this species isolated from *G. geographica*,

G. pseudogeographica, and *K. f. spooneri* were morphologically indistinguishable from those of *E. mitraria* isolated from *C. p. belli*, and therefore were considered to be those of *E. mitraria* (Wacha & Christiansen, 1976). Oocysts from *G. geographica* were 11.0 × 8.3 (9−12 × 8−9), L/W 1.3 (1.1−1.5), and sporocysts were 6.0 × 3.8 (5−7 × 3−4); those from *G. pseudogeographica* were 9.2 × 8.1 (8−11 × 7−9), L/W 1.1 (1.0−1.5), and sporocysts were 6.6 × 3.8 (3−4); and those from *K. f. spooneri* were 11.5 × 8.6 (10−13 × 7−10), L/W 1.3 (1.3−1.6), and sporocysts were 6.1 × 3.8 (5−8 × 3−4) (Wacha & Christiansen, 1976). Široký and Modrý (2006) found oocysts indistinguishable from *E. mitraria* in 9/9 (100%) *H. depressa*, which had ovoidal oocysts 14.9 × 10.1 (13−16 × 9−12) with L/W 1.5 (1.3−1.8) and ovoidal to ellipsoidal sporocysts 6.8 × 4.2 (6−8 × 3.5−5) with L/W 1.6 (1.3−2.0). Segade et al. (2006) found oocysts like these in 11/44 (25%) *E. orbicularis* from northwestern Spain; the oocysts they saw were miter-shaped with one conical projection at one end and two or three at the other end and were 12.1 × 9.5 (10−14 × 7−11) with L/W 1.3 (1.1−1.5), and ellipsoidal sporocysts that were 4.5 × 3.3 (3−6 × 3−4), with L/W 1.3 (1.0−1.7), a small SB, and a spheroidal OR that was membrane-bound. Finally, Široký et al. (2006a) also measured and described oocysts from *M. heliostemma* (Amazon Toadhead Turtles); the oocysts they studied were 14.4 × 10.4 (13−16 × 9−13), L/W 1.4 (1.2−1.6), lacking an M, OR, and PG. One end of the oocysts they saw had one conical projection, ~1.0−1.5 long, while the opposite end had three similar projections. The ellipsoidal sporocysts they measured were 7.1 × 4.4 (6−7.5 × 4−5), L/W 1.6 (1.3−1.9), with a tiny, knob-like SB, but no SSB or PSB and the SR was a compact granular mass ~2−4 wide.

This species presents a bit of an enigma because the morphotype reported by everyone seems to be in multiple families of turtles. Although McAllister et al. (1990) mention that it was reported from two families, they try to cast doubt on the report that found it in a third family, the Kinosternidae. Segade et al. (2006) summarized this species as a "ubiquitous euryxenic eimerian having been found in nine emydid, two kinosternid, and one chelydrid turtles, all of them, excepting the Asiatic Reeve's turtle *Chinemys reevisii*, from North America," when they gave their first report of this eimerian in European emydid turtles. Clearly, molecular data is desperately needed to sort out this mess.

Genus *Melanochelys* Gray, 1834 (2 Species)
Eimeria patta Široky & Modrý, 2005

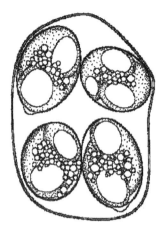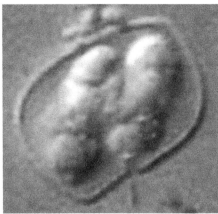

Figures 2.81, 2.82 Line drawing of the sporulated oocyst of Eimeria patta *from Široky and Modrý (2005), with permission from the journal* Parasite *and from both authors. Photomicrograph of a sporulated oocyst of* E. patta *from Široky and Modrý (2005), with permission from the journal* Parasite *and from both authors.*

Type host: *Melanochelys trijuga* (Schweigger, 1802), Indian Black Turtle of Indian Pond Terrapin.

Type locality: ASIA: Myanmar (Burma): detailed collection locality unknown.

Other hosts: None to date.

Geographic distribution: ASIA: Myanmar (Burma).

Description of sporulated oocyst: Oocyst shape: irregular as they range from ovoidal, ellipsoidal to almost subspheroidal; number of walls: 1; wall characteristics: smooth, colorless, very thin, ~0.2−0.3 thick; L × W: 12.6 × 9.1 (11−16 × 7.5−12); L/W ratio: 1.4 (1.1−1.6); M, OR, PG: all absent. Distinctive features of oocyst: very thin wall that makes the shape highly variable.

Description of sporocyst and sporozoites: Sporocyst shape: ovoidal to ellipsoidal; L × W: 5.8 × 4.2 (5−7 × 3.5−5); L/W ratio: 1.4 (1.2−1.75); SB: present, a flat projection at one end of the sporocyst; SSB, PSB: both absent; SR: present; SR characteristics: granular matter scattered among SZ; SZ: encircle each other and lie head-to-tail (sometimes curved together), each with a subspheroidal RB at both ends. Distinctive features of sporocyst: small flat SB, disbursed SR of tiny granules, and SZ with a RB at each end.

Prevalence: Unknown. Examined samples originated from a group of three turtles.

Sporulation: Unknown. Collected fresh feces were placed in 2.5% (w/v) $K_2Cr_2O_7$ solution and "allowed to sporulate at room temperature" (°C, not stated), but these methods did not allow Široky and Modrý (2010) to determine if the oocysts were unsporulated when first placed into 2.5% (w/v) $K_2Cr_2O_7$.

Endogenous stages: Unknown. Oocysts were collected from the feces of live animals.

Cross-transmission: None to date.

Pathology: Unknown.

Material deposited: Photosyntypes deposited in the Department of Parasitology, University of Veterinary and Pharmaceutical Sciences Brno, Czech Republic, No. R 78/04.

Entymology: "Leik patta" is one of the vernacular names for the type host in Myanmar; the specific epithet "*patta*" was given in accordance with the International Code of Zoological Nomenclature (Article 31.1) as a noun in apposition (ICZB, 1999).

Remarks: Široky and Modrý (2005) mentioned that host systematics and geographic origin are commonly used criteria in making taxonomic decisions regarding the identification of members of the genus *Eimeria*. Thus, in making their decision to name this form a new species only *Eimeria* species described and named from turtles of the families Emydidae, Geoemydidae, and Testudinidae (Superfamily Testudinoidea) were used for comparative purposes, "as conspecificity of described species with coccidia from other hosts is unlikely." Sporulated oocysts of *E. patta*, according to the authors (Široky & Modrý, 2005), has a uniquely thin oocyst wall when compared to other *Eimeria* species from chelonian hosts. Its markedly irregular shape of oocysts distinguishes this species from all other eimerians except those of *E. motelo*, a species from the Neotropical yellow-footed tortoise; however, both oocysts and sporocysts of the latter species are markedly longer.

Genus *Morenia* Duméril & Bibron, 1835 (2 Species)

To our knowledge, there are no coccidia described from this genus.

Genus *Notochelys* Gray, 1834 (Monospecific)

To our knowledge, there are no coccidia described from this genus.

Genus *Orlitia* Gray, 1873 (Monospecific)

To our knowledge, there are no coccidia described from this genus.

Genus *Pangshura* Gray, 1831 (4 Species)

Eimeria pangshurae Široký & Modrý, 2010

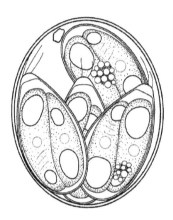

Figures 2.83, 2.84 Line drawing of the sporulated oocyst of Eimeria pangshurae *from Široký and Modrý (2010), with permission from* Acta Protozoologica *and from both authors. Photomicrograph of a sporulated oocyst of* E. pangshurae *from Široký and Modrý (2010), with permission from* Acta Protozoologica *and from both authors.*

Type host: *Pangshura sylhetensis* Jerdon, 1870, Assam Roofed Turtle.

Type locality: ASIA: India: Assam.

Other hosts: None to date.

Geographic distribution: ASIA: India.

Description of sporulated oocyst: Oocyst shape: spheroidal to ellipsoidal; number of walls: 1; wall characteristics: smooth, colorless, 1.0 thick; L × W (N = 20): 16.5 × 13.2 (15−17 × 10−17); L/W ratio: 1.3 (1.0−1.6); M, OR: both absent; PG: present, a single, spheroidal to subspheroidal, 1.5−2 wide. Distinctive features of oocyst: spheroidal with a distinct PG, but lacking both an M and an OR.

Description of sporocyst and sporozoites: Sporocyst shape: broadly ellipsoidal to flask-shaped; L × W: 11.1 × 5.7 (10−12 × 5−7); L/W ratio: 2.0 (1.6−2.2); SB: present, a flat projection usually connected with 2−4 filamentous structures; SSB, PSB: both absent; SR: present; SR characteristics: small body composed of tiny granules sometimes scattered among SZ; SZ: banana-shaped (line drawing), lie head-to-tail (sometimes curved together), each with a larger sub-spheroidal RB at rounded end and a smaller one at pointed end with a small N (1−1.5 wide) located between them in center of SZ. Distinctive features of sporocyst: distinct flat SB and a disbursed SR of tiny granules.

Prevalence: Unknown. Examined samples originated from a group of three turtles.

Sporulation: Unknown. Collected fresh feces were placed in 2.5% (w/v) $K_2Cr_2O_7$ solution and "allowed to sporulate at room temperature" (°C, not stated), but these methods did not allow Široky and Modrý (2010) to determine if the oocysts were unsporulated when first placed into 2.5% (w/v) $K_2Cr_2O_7$.

Endogenous stages: Unknown. Oocysts were collected from the feces of live animals.

Cross-transmission: None to date.

Pathology: Unknown.

Material deposited: Photosyntypes deposited in the collection of the Institute of Parasitology, Biology Centre of the Academy of Sciences of the Czech Republic, České Budějovice, Czech Republic, No. IP ProtColl P10.

Entymology: The specific epithet was derived as a genitive of the host genus, which is grammatically a feminine name.

Remarks: Sporulated oocysts of this species somewhat resemble those of five previously described eimerians. They somewhat resemble those of *E. carri* and *E. juniataensis*, but these both have an OR which those of *E. pangshurae* lack. They somewhat resemble those of *E. galaecianensis*, but its oocysts also have an OR and are larger than those of *E. pangshurae* (19.3 × 16 vs. 16.5 × 13.2) and the SB of the two have different shapes. Sporulated oocysts of *E. megalostiedae* are similar in shape to those of *E. pangshurae*, but are smaller (13.9 × 12.8 vs. 16.5 × 13.2), have a membrane-bound OR that *E. pangshurae* lacks, and have a large, elongated SB that lacks the filamentous projections found on sporocysts of *E. pangshurae*. Finally, sporulated oocysts of *E. ornata* are bigger (17.9 × 15.7 vs. 16.5 × 13.2) and have an OR that those of *E. pangshurae* lack.

Eimeria kachua Široky & Modrý, 2005

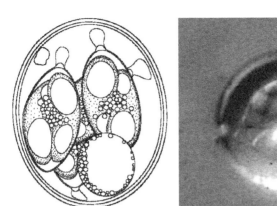

Figures 2.85, 2.86 Line drawing of the sporulated oocyst of Eimeria kachua *from Široky and Modrý (2005), with permission from the journal* Parasite *and from both authors. Photomicrograph of a sporulated oocyst of* E. kachua *from Široky and Modrý (2005), with permission from the journal* Parasite *and from both authors.*

Type host: *Pangshura tentoria circumdata* (Martens, 1969) (syn. *Kachuga tentoria circumdata* Martens, 1989), Pink-Ringed or Indian Tent Turtle.

Type locality: ASIA: India: at a market in Bombay, but the details of the collection locality are unknown.

Other hosts: None to date.

Geographic distribution: ASIA: India.

Description of sporulated oocyst: Oocyst shape: subspheroidal to broadly ovoidal; number of walls: 2; wall characteristics: smooth, colorless, fragile, extraordinarily thin, ~0.8 (0.7−0.9) thick; L × W: 15.3 × 13.9 (13−18 × 12−16); L/W ratio: 1.1 (1.0−1.2); M: absent; OR: present; OR characteristics: composed of fine granules and covered with a very thin membranous layer, 5.0 × 5.5 (3.5−6 × 4−7); PG: present, usually a single, small granule, 1.0−1.5 × 1.2−1.7, but rarely there are two granules. Distinctive features of oocyst: small, subspheroidal with a distinct PG (sometimes two) and an OR.

Description of sporocyst and sporozoites: Sporocyst shape: elongate-ovoidal to spindle-shaped; L × W: 8.7 × 4.9 (7.5−10 × 4−6); L/W ratio: 1.8 (1.5−2.0); SB: present, knob-like, 0.5−0.8 high × 1.2−1.5 wide and covered with a fine, membranous cupola-like structure, 1.2−1.8 wide × 1.8−2.0 high; SSB, PSB: both absent; SR: present; SR characteristics: a small amount of tiny granules scattered among

SZ; SZ: comma-shaped (line drawing), lie head-to-tail (sometimes curved together), each with a large subspheroidal to ovoidal RB, $2-2.5 \times 2-3$, at rounded end and a smaller one, $\sim 1.5-2$, near the more pointed end; N of the SZ was not discernable. Distinctive features of sporocyst: membranous cupola-like structure that covers the SB.

Prevalence: Unknown. Samples examined by Široky and Modrý (2005) originated from a group of 10 turtles.

Sporulation: Exogenous. Fresh feces were placed in 2.5% (w/v) potassium dichromate ($K_2Cr_2O_7$) solution at $20-23°C$ and some oocysts became fully sporulated within $3-5$ days; however, not all oocysts sporulated by the end of 30 days in the dichromate solution.

Prepatent and patent periods: Unknown.

Site of infection: Unknown. Fecal samples were obtained from live turtles.

Endogenous stages: Unknown. Oocysts were collected from the feces of live animals.

Cross-transmission: None to date.

Pathology: Unknown.

Material deposited: Photosyntypes are deposited in the Department of Parasitology, University of Veterinary and Pharmaceutical Sciences, Brno, Czech Republic, No. R 79/04.

Entymology: The specific epithet "*kachua*" was adopted from Hindi language, meaning a turtle. It was given as a noun in apposition, in accordance with the International Code of Zoological Nomenclature (Article 31.1, ICZN, 1999).

Remarks: Široky and Modrý (2005) mentioned that host systematics and geographic origin are commonly used criteria in making taxonomic decisions regarding the identification of members of the genus *Eimeria*. Thus, in making their decision to name this form a new species, only *Eimeria* species described and named from turtles of the families Emydidae, Geoemydidae, and Testudinidae (Superfamily Testudinoidea) were used for comparative purposes, "as conspecificity of described species with coccidia from other hosts is unlikely." The sporulated oocysts of this form are similar in size and shape to those of six other *Eimeria* from testudinoids, including *E. carri*, *E. graptemydos*, *E. jaboti*, *E. juniataensis*, *E. megalostiedai*, and *E. ornata*. Oocysts of *E. carri* are different by having longer sporocysts, oocysts without a PG, and only a tiny SB on the sporocysts. Oocysts of *E. graptemydos* are slightly smaller with a single-layered and very thin oocyst wall. Oocysts of *E. jaboti* are somewhat bigger ($17-19 \times 15-17$ vs. $13-18 \times 12-16$), lack an OR, and have strictly ovoidal sporocysts that lack an SB. Oocysts of *E. juniataensis* are smaller, mostly spheroidal with a very thin wall composed of one layer, and lack a PG, which those of *E. kachua* possess. Oocysts of *E. megalostiedai* also have a thinner oocyst wall, lack a PG, and have sporocysts with very large SBs. Oocysts of *E. ornata* are said to mostly sporulate endogenously vs. the mostly exogenous sporulation of *kachua* and, in addition, this form differs markedly from all other species by having a fine membranous structure covering its SB.

Genus *Rhinoclemmys* Fitzinger, 1835 (9 Species)

To our knowledge, there are no coccidia described from this genus.

Genus *Sacalia* Gray, 1870 (2 Species)

To our knowledge, there are no coccidia described from this genus.

Genus *Siebenrockiella* Lindholm, 1929 (2 Species)

To our knowledge, there are no coccidia described from this genus.

Genus *Vijayachelys* Henderson, 1912 (Monospecific)

To our knowledge, there are no coccidia described from this genus.

FAMILY PLATYSTERNIDAE, BIG-HEADED TURTLES, 1 GENUS, MONOTYPIC

Genus *Platysternon* Gray, 1831 (Monospecific)

To our knowledge, there are no coccidia described from this genus.

SUPERFAMILY TRIONYCHOIDEA

FAMILY CARETTOCHELYIDAE, PIGNOSE TURTLES, 1 GENUS, MONOTYPIC

Genus *Carettochelys* Ramsay, 1886 (Monospecific)

To our knowledge, there are no coccidia described from this genus.

FAMILY TRIONYCHIDAE, SOFTSHELL TURTLES, 13 GENERA, 30 SPECIES

Genus *Amyda* Saint-Hilaire, 1809 (Monospecific)

To our knowledge, there are no coccidia described from this genus.

Genus *Apalone* Rafinesque, 1832 (3 Species)
Eimeria amydae Roudabush, 1937

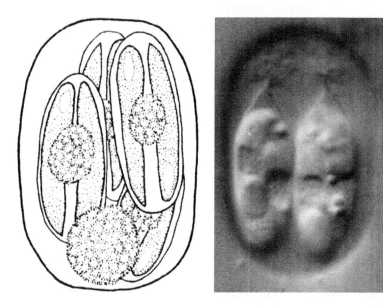

Figures 2.87, 2.88 Line drawing of the sporulated oocyst of Eimeria amydae *from Roudabush (1937), with permission from the* Journal of Parasitology. *Photomicrograph of a sporulated oocyst of* Eimeria amydae *from McAllister et al. (1990c), with permission from the* Journal of Parasitology *and from the senior author.*

Type host: *Apalone spinifera* (Le Sueur, 1827) (syn. *Amyda spinifera*, Burt, 1935), Eastern Spiny Softshell Turtle.

Type locality: NORTH AMERICA: USA: Iowa, Ames.

Other hosts: *Apalone spinifera pallida* (Webb, 1962), Pallid Spiny Softshell Turtle.

Geographic distribution: NORTH AMERICA: USA: Iowa, Texas.

Description of sporulated oocyst: Oocyst shape: ovoidal (Roudabush, 1937) to elongate-ellipsoidal (McAllister et al., 1990); number of walls: 1; wall characteristics: very thin; L × W: 19.6 × 14.6 (17−24 × 12−17); L/W ratio: 1.3; M, PG: both absent; OR: present as a large, spheroidal mass composed of small granules (original line drawing). Distinctive features of oocyst: very thin wall and large OR.

Description of sporocyst and sporozoites: Sporocyst shape: ellipsoidal, described as ⅔ the length of the oocyst; SB: absent according to Roudabush (1937, Figure 4), but present with a unique stalk with attached filaments (see Figures 11, 12 in McAllister et al., 1990); SSB, PSB: both absent; SR: present; SR characteristics: a spheroidal mass of small granules in the middle of the sporocyst, between the SZ; SZ: elongate sausage-shaped without an RB (Roudabush, 1937, Figure. 4), but more likely with two distinct RBs (McAllister et al., 1990, see *Remarks* below). Distinctive features of sporocyst: hot dog-shaped SZ and a spheroidal SR of small granules (but see more unique descriptive features provided by McAllister et al., 1990, under *Remarks* below).

Prevalence: In 1/1 (100%) of type host from Iowa; in 1/10 (10%) *A. s. pallida* from Texas (McAllister et al., 1994).

Sporulation: Exogenous. Oocysts became fully sporulated in 18 h when placed in 2% aqueous $K_2Cr_2O_7$ (Roudabush, 1937) or in 48−72 h at 23°C in 1% (W/V) aqueous $K_2Cr_2O_7$ (McAllister et al., 1990).

Prepatent and patent periods: Unknown. Oocysts collected from intestinal contents.

Site of infection: Roudabush (1937) reported he found oocysts in both the small and large intestines of a softshell turtle, but not in the gall bladder or bile ducts.

Endogenous stages: Unknown.

Cross-transmission: None to date.

Pathology: None reported.

Materials deposited: None.

Remarks: Roudabush (1937) described and named this species from a turtle in Iowa and the description of the oocyst and sporocysts above is from his work unless noted otherwise. McAllister et al. (1990) also found this species in turtles in Texas; their oocysts were 21.8×14.6 ($18−24 \times 14−16$), L/W ratio 1.5 (1.3−1.6), with a spheroidal OR that was essentially a large vacuole, 8.7×8.5 ($7−10 \times 7−10$); an M was absent and a PG was rarely present. Roudabush (1937) did not give measurements for sporocysts, but McAllister et al. (1990) did. The sporocysts they measured were elongate-ovoidal, 14.2×5.8 ($12−16 \times 5−6$), L/W ratio 2.5 (2.1−2.8) with an SB that had delicate, thin filaments 4−5 long that arose from a stalk at the tip of the SB, but not the SB itself; an SSB and PSB were both absent. The sporocysts had an SR usually of scattered granules, but sometimes as a cluster of small granules; SZ were 12.3×2.9 ($10-14 \times 2-3$) with transverse striations at their pointed end and an anterior RB that was 2.3×2.4 ($2-3 \times 2-3$) and a larger posterior RB that was 2.6×3.8 ($2-3 \times 3-5$) (McAllister et al., 1990). McAllister et al. (1990) also included a photomicrograph of a sporulated oocyst which the original description by Roudabush (1937) lacked.

Eimeria apalone McAllister, Upton, & McCaskill, 1990a

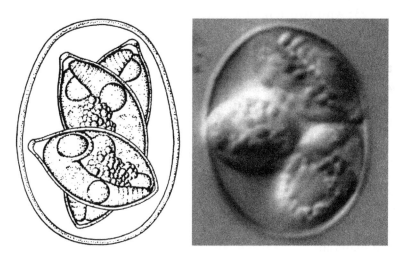

Figures 2.89, 2.90 Line drawing of the sporulated oocyst of Eimeria apalone from McAllister et al. (1990a), with permission from the Journal of Parasitology and from the senior author. Photomicrograph of a sporulated oocyst of E. apalone from McAllister et al. (1990a), with permission from the Journal of Parasitology and from the senior author.

Type host: *Apalone spinifera pallida* (Webb, 1962), Pallid Spiny Softshell Turtle.

Type locality: NORTH AMERICA: USA: Texas, Dallas County, Dallas, Cedar Creek at Dallas Zoo, 621 East Clarendon Drive.

Other hosts: *Apalone spinifera hartwegi* (Conant & Goin, 1941), Western Spiny Softshell Turtle.

Geographic distribution: NORTH AMERICA: USA: Arkansas, Texas.

Description of sporulated oocyst: Oocyst shape: ellipsoidal, elongate pear-shaped, or rarely subspheroidal; number of walls: 1; wall characteristics: membrane-like, thin, apparently single-layered, ~ 0.4

thick; $L \times W$ (N = 25): 16.8×13.2 $(12-19 \times 10-16)$; L/W ratio: 1.3 $(1.0-1.5)$; M, OR, PG: all absent. Distinctive features of oocyst: very thin oocyst wall that sometimes collapses, adhering tightly to enclose sporocysts.

Description of sporocyst and sporozoites: Sporocyst shape: elongate-ovoidal; $L \times W$ (N = 25): 11.3×6.2 $(9-14 \times 5-7)$; L/W ratio: 1.8 $(1.6-2.2)$; SB: present, ~ 1.6 wide $\times 1.0$ high; SSB, PSB: both absent; SR: present; SR characteristics (N = 23): spheroidal to subspheroidal, 4.7×3.8 $(2-7 \times 2-3)$, composed of cluster of large coarse granules; SZ (N = 23): elongate, 12.9×2.7 $(10-14 \times 2-3)$, with anterior ends sometimes bearing transverse striations; each SZ (N = 23) with two RBs; the anterior one is spheroidal to subspheroidal, 2.3×2.4 $(2-3 \times 2-4)$ and the posterior one is ovoidal to subspheroidal, 2.5×3.6 $(2-3 \times 2-4)$; N not easily seen, but located between RBs. Distinctive features of sporocyst: prominent SB and SZ each with two RBs.

Prevalence: In 5/5 (100%) of the type host from Dallas County, but in 0/2 (0%) from Johnson County and 0/2 (0%) from Somervell County, Texas; later, found in 1/3 (33%) *A. s. hartwegi* from Arkansas and in 5/10 (50%) *A. s. pallida*, also from Texas (McAllister et al., 1994).

Sporulation: Exogenous. Oocysts became fully sporulated after 72 h at 23°C in 1% $K_2Cr_2O_7$.

Prepatent and patent periods: Unknown. Oocysts collected from fecal material.

Site of infection: Unknown. Oocysts found in feces.

Endogenous stages: Unknown.

Cross-transmission: None to date.

Pathology: Unknown.

Materials deposited: Syntypes (oocysts in 10% formalin) deposited in the US National Parasite Collection, Beltsville, Maryland, USNPC No. 80748.

Entymology: The specific epithet is derived from the word *apalos* (Greek for "soft" or "tender") to describe the shell of the turtle type host.

Remarks: Sporulated oocysts of this species most closely resemble those of *E. spinifera* in size. However, *E. apalone* does not possess an SB bearing filaments, an OR, or a PG. Further, it does not closely resemble any of the other species of turtle eimerians described to date.

Eimeria dericksoni Roudabush, 1937

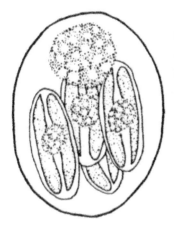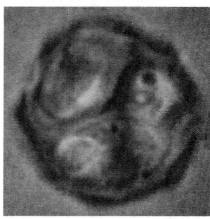

Figures 2.91, 2.92 Line drawing of the sporulated oocyst of Eimeria dericksoni *from Roudabush (1937), with permission from the* Journal of Parasitology. *Photomicrograph of a sporulated oocyst of* E. dericksoni *from Wacha and Christiansen (1977), with permission from John Wiley & Sons, Ltd. holder of the copyright for the* Journal of Eukaryotic Microbiology *(formerly* Journal of Protozoology*).*

Type host: *Apalone spinifera* (Le Sueur, 1827) (syn. *Amyda spinifera*, Burt, 1935), Eastern Spiny Softshell Turtle.
Type locality: NORTH AMERICA: USA: Iowa, Ames.
Other hosts: *Apalone spinifera hartwegi* (Conant & Goin, 1941), Western Spiny Softshell Turtle; *Apalone spinifera pallida* (Webb, 1962), Pallid Spiny Softshell Turtle.
Geographic distribution: NORTH AMERICA: USA: Iowa, Texas.
Description of sporulated oocyst: Oocyst shape: usually subspheroidal; number of walls: 1; wall characteristics: composed of only one layer, very thin; L × W: 14.6 × 12.9 (12−17 × 11−16); L/W ratio: 1.1; M: absent; PG: either absent (Roudabush, 1937) or present (Wacha & Christiansen, 1977); OR: present as a large spheroidal body composed of small granules (line drawing). Distinctive features of oocyst: very thin wall and OR thicker than the sporocysts.
Description of sporocyst and sporozoites: Sporocyst shape: ellipsoidal, described as ½ the length of the oocyst; SB: said to be present on some sporocysts, but were not shown in original line drawing; SSB, PSB, both absent; SR: present; SR characteristics: a large spheroidal mass of small granules located at one end of sporocyst (line drawing); SZ: hot dog-shaped, with no discernable RB or N (line drawing).

Prevalence: In 1/1 (100%) of the type host; in 3/7 (43%) *A. s. pallida* from Texas including 2/5 (40%) from Dallas County and 1/2 (50%) from Johnson County; later, it was found in 3/10 (30%) *A. s. pallida* also in Texas (McAllister et al., 1994).

Sporulation: According to Roudabush (1937) oocysts sporulated in slightly >24 h when placed in 2% aqueous $K_2Cr_2O_7$.

Prepatent and patent periods: Unknown. Oocysts collected from intestinal contents.

Site of infection: Roudabush (1937) reported that he found oocysts in both the small and large intestines of a softshell turtle, but not in the gall bladder or bile ducts.

Endogenous stages: Unknown.

Cross-transmission: None to date.

Pathology: None reported.

Materials deposited: None.

Entymology: This species was named after Dr. S.H. Derickson of Lebanon Valley College, whose encouragement and diligent teaching was Roudabush's incentive for beginning his scientific career.

Remarks: Roudabush (1937) described this species in a turtle he identified as *Amyda spinifera*, but that name was changed to *Apalone spinifera*, as noted above (Uetz & Hošek, 2013). Wacha and Christiansen (1976) described six new coccidian species from turtles collected in Iowa, and they made mention of *E. dericksoni*, but only in a manner that is more confusing than insightful. They stated, "Collection data for the infected hosts are summarized separately…" and "…Differential diagnoses having reference to …*E. dericksoni* Roudabush, 1937, in the following portions of the text, are based, in part, on our observations of oocysts of these species which were isolated by us from the type host, *Trionyx spiniferus* La Sueur. Therefore, some of the data concerning these 2 species are new." However, they never mention how many of the five turtles sampled were infected with this eimerian, nor do they indicate any new data. For example, when comparing the sporocyst SBs of *E. graptemydos* to those of *E. dericksoni*, Wacha and Christiansen (1976) only said, "confirmed by direct observation of the oocysts of *E. dericksoni*."

In their second paper on coccidians from turtles in Iowa (Wacha & Christiansen, 1977), they redescribed *E. dericksoni* from *A. s. hartwegi*, and added photomicrographs and a line drawing of the sporulated oocyst. It should be noted that their redescribed oocysts were smaller than those in the original description (used above), i.e., 10.8 × 10.0 (9–13 × 8.5–12) vs. 14.6 × 12.9 (12–17 × 11–16), but the oocyst L/W ratios (1.1) are the same. Their redescription (1977) also included structural aspects of the SB, dimensions of sporocysts and SR, and the presence of a PG, which Roudabush (1937) did not report in his original description. McAllister et al. (1990) also reported this species in *A. s. pallidus* from Texas.

Eimeria mascoutini Wacha & Christiansen, 1976

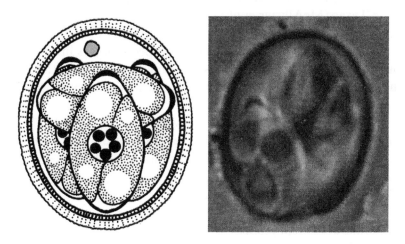

Figures 2.93, 2.94 *Line drawing of the sporulated oocyst of* Eimeria mascoutini *(original) adapted from Wacha and Christiansen (1976). Photomicrograph of a sporulated oocyst of* E. mascoutini *from Wacha and Christiansen (1977), with permission from John Wiley & Sons, Ltd. holder of the copyright for the* Journal of Eukaryotic Microbiology *(formerly* Journal of Protozoology*).*

Type host: *Apalone spinifera* (La Sueur, 1827) (syn. *Trionyx spiniferus* [Le Sueur]), Spiny Softshell Turtle.
Type locality: NORTH AMERICA: USA: Iowa.
Other hosts: *Apalone spinifera pallida* (Webb, 1962), Pallid Spiny Softshell Turtle.
Geographic distribution: NORTH AMERICA: USA: Iowa.
Description of sporulated oocyst: Oocyst shape: broadly ellipsoidal to subspheroidal; number of walls: 1; wall characteristics: outer surface mammillated, with outer portion striated in cross section, stippled in tangential section, ~0.6–1.0 thick; L × W: 14.0 × 11.9 (11.5–16 × 10–14); L/W ratio: 1.2 (1.1–1.3); M, OR: both absent;

PG: present. Distinctive features of oocyst: mammillated outer wall and presence of PG while lacking an OR.

Description of sporocyst and sporozoites: Sporocyst shape: narrowly ellipsoidal; L × W: 9.6 × 5.3 (9−11.5 × 4.5−6); L/W ratio: 1.8; SB: present, thin, convex; SSB, PSB: both absent; SR: present; SR characteristics: spheroidal membrane-bound body, 1.9−3.8, with 3−6 spheroidal granules, each ~0.6; SZ: one large ellipsoidal to spheroidal RB at the broad end and a small spheroidal RB at narrow end. Distinctive features of sporocyst: spheroidal shape, membrane-bounded SR and SZs each with two RBs.

Prevalence: In 2/5 (40%) of the type host; in 3/10 (30%) *A. s. pallida* from Texas (McAllister et al., 1994).

Sporulation: Likely exogenous but unknown because Wacha and Christiansen (1977) put turtle feces into specimen jars in 2.5% aqueous (w/v) $K_2Cr_2O_7$ solution at ~22°C for 1 week before they examined them for oocysts. Samples with sporulated oocysts were then stored in a refrigerator at ~5°C.

Prepatent and patent periods: Unknown.

Site of infection: Likely the intestines because oocysts were found only in the feces from the intestine and not in the bile.

Endogenous stages: Unknown.

Cross-transmission: None to date.

Pathology: Unknown.

Materials deposited: None.

Entymology: The specific epithet of this eimerian is coined for the Mascoutin Indian tribe, after which the Iowa county of Muscatine, the type locality for this species, is named.

Remarks: Following the original description of this species in which they only included a line drawing of the sporulated oocyst (Wacha & Christiansen, 1976), Wacha and Christiansen (1977) later added photomicrographs of the oocysts and added additional measurements of the oocysts and sporocysts. Newly measured (1977) sporulated oocysts (N = 43) were 13.8 × 12.0 (11.5−16.5 × 10−14), L/W 1.2 (1.1−1.4), and sporocysts (N = 50) were 8.6 × 5.3 (7−11 × 4.5−6.5), L/W 1.6.

Two other eimerians from turtles have sporulated oocysts which, like those of *E. mascoutini*, are broadly ellipsoidal to subspheroidal, have a PG, lack an OR, and enclose ellipsoidal sporocysts. Those of *E. mascoutini* differ from *E. jaboti* by having smaller oocysts (11.5–16 × 10–14 vs. 17–19 × 15–17) and by having sporocysts with SBs, which those of *E. jaboti* lack; and they differ from *E. paynei* by having smaller oocysts (14 × 12 vs. 23 × 19) and shorter sporocysts (9.6 vs. 13.2).

Eimeria pallidus McAllister, Upton, & McCaskill, 1990a

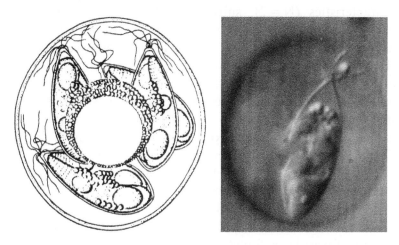

Figures 2.95, 2.96 Line drawing of the sporulated oocyst of Eimeria pallidus *from McAllister et al. (1990a), with permission from the* Journal of Parasitology *and the senior author. Photomicrograph of a sporulated oocyst of* E. pallidus *from McAllister et al. (1990a), with permission from the* Journal of Parasitology *and the senior author.*

Type host: *Apalone spinifera pallida* (Webb, 1962), Pallid Spiny Softshell Turtle.
Type locality: NORTH AMERICA: USA: Texas, Dallas County, Dallas, Cedar Creek at Dallas Zoo, 621 East Clarendon Drive.
Other hosts: None to date.
Geographic distribution: NORTH AMERICA: USA: Texas.
Description of sporulated oocyst: Oocyst shape: spheroidal or sub-spheroidal, with one side often slightly narrower; number of walls: 1; wall characteristics: smooth, thin, apparently single-layered, ~0.5 thick; L × W (N = 25): 23.4 × 21.6 (18–27 × 17–25); L/W ratio: 1.1 (1.0–1.3); M: absent; OR: present as spheroidal or sub-spheroidal globule, 10.9 × 10.6 (6–14 × 6–14), surrounded by small granules that appear to coalesce on one side (line drawing);

PG: present in only 20% of oocysts. Distinctive features of oocyst: large OR composed of a large globule surrounded by small granules on one side.

Description of sporocyst and sporozoites: Sporocyst shape: elongate-ovoidal; $L \times W$ (N = 20): 14.3×6.2 ($13-17 \times 6-7$); L/W ratio: 2.3 (2.0–2.7); SB: present, as a thickening of one end of sporocyst and bearing six to ten short filaments 4–5 long; SSB, PSB: both absent; SR: occasionally present; SR characteristics (N = 2): subspheroidal or ovoidal, 4.4×4.0 or more often as a loose cluster of granules or as granules scattered among SZ; SZ (N = 16): elongate, 12.5×2.9 ($11-14 \times 2-3$), usually lying in opposite directions, but sometimes facing same way within sporocyst; anterior ends of SZ bear six to eight transverse striations; each SZ (N = 16) with spheroidal or subspheroidal anterior RB, 2.2×2.3 ($2-3 \times 2-3$) and subspherioidal or ovoidal posterior RB, 2.7×3.2 ($2-3 \times 2-4$); N lies between RBs. Distinctive features of sporocyst: SB with six to ten filaments and SZ with six to eight transverse striations at anterior end.

Prevalence: In 4/5 (80%) of the type host from Dallas County, but in 0/2 (0%) from Johnson County and 0/2 (0%) from Somervell County, Texas; later, found in 4/10 (40%) of the same host species, also from Texas (McAllister et al., 1994).

Sporulation: Exogenous. Oocysts became fully sporulated after 72 h at $\sim 23°C$ in 1% (w/v) $K_2Cr_2O_7$.

Prepatent and patent periods: Unknown.

Site of infection: Unknown. Oocysts found in feces.

Endogenous stages: Unknown.

Cross-transmission: None to date.

Pathology: Unknown.

Materials deposited: Syntypes (oocysts in 10% formalin) deposited in the US National Parasite Collection, Beltsville, Maryland, USNPC No. 80758.

Entymology: The specific epithet is derived from the word *pallid* (Latin for "pale") to describe the juvenile pattern of the type host.

Remarks: Sporulated oocysts of this species are similar to those of *E. spinifera*, but this species has much larger oocysts and more elongate sporocysts.

Eimeria spinifera McAllister, Upton, & McCaskill, 1990a

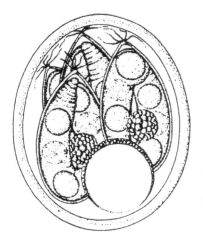

Figures 2.97, 2.98 Line drawing of the sporulated oocyst of Eimeria spinifera *from McAllister et al. (1990a), with permission from the* Journal of Parasitology *and the senior author. Photomicrograph of a sporulated oocyst of* E. spinifera *from McAllister et al. (1990a), with permission from the* Journal of Parasitology *and the senior author.*

Type host: *Apalone spinifera pallida* (Webb, 1962), Pallid Spiny Softshell Turtle.

Type locality: NORTH AMERICA: USA: Texas, Somervell County, 14.5 km NW Glen Rose off county road 308 at Georges Creek.

Other hosts: None to date.

Geographic distribution: NORTH AMERICA: USA: Texas.

Description of sporulated oocyst: Oocyst shape: subspheroidal, ellipsoidal, or occasionally pear-shaped; number of walls: 1; wall characteristics: smooth, thin, apparently single-layered wall, ~0.8 thick; L × W (N = 25): 16.3 × 14.0 (14−19 × 12−18); L/W ratio: 1.2 (1.1−1.3); M: absent; OR: present; OR characteristics: spheroidal, homogenous globule, 6.3 × 6.3 (5−8 × 5−8) and sometimes a smaller globule, ~1.8, occurs within larger globule; PG: present in 16% of oocysts, attached to OR as a cluster of small granules aggregated along one side of OR and may reach 4−6 wide. Distinctive features of oocyst: PG of small granules that aggregate on one side of OR.

Description of sporocyst and sporozoites: Sporocyst shape: elongate-ovoidal; L × W: 10.3 × 5.2 (8−12 × 5−6); L/W ratio: 2.0 (1.8−2.2); SB: present, consisting of small knob-like thickening bearing 4−5 short filaments ≤4 long; SSB, PSB: both absent; SR: present; SR characteristics: compact spheroidal or subspheroidal, granular,

2.5 × 2.4 (1–3 × 1–3), rarely scattered among SZ; SZ (N = 18): elongate, 9.4 × 2.6 (9–10 × 2–3) *in situ*, arranged head-to-tail with posterior ends recurved back along one end. Anterior end of each SZ bears transverse striations and each contains spheroidal anterior RB (N = 18), 2.1 × 2.1 (1.6–2.4 × 1.6–2.4) and a spheroidal or sub-spheroidal posterior RB (N = 18), 2.2 × 2.3 (2–3 × 2–3); N between two RBs. Distinctive features of sporocyst: SB with filaments.

Prevalence: In 1/2 (50%) of the type host from the type locality; in 2/5 (40%) of the same host from Dallas County and 0/2 (0%) from Johnson County, Texas; later, found in 3/10 (30%) of the same host species, also in Texas (McAllister et al., 1994).

Sporulation: Exogenous. Oocysts sporulated in 48–72 h at 23°C in tap water supplemented with antibiotics.

Prepatent and patent periods: Unknown. Oocysts recovered from fecal material.

Site of infection: Unknown. Oocysts found in fecal material from intestinal contents, but not bile.

Endogenous stages: Unknown.

Cross-transmission: None to date.

Pathology: Unknown.

Materials deposited: Syntypes (oocysts in 10% formalin) deposited in the US National Parasite Collection, Beltsville, Maryland, USNPC No. 80747. Symbiotype host in the Arkansas State University Museum of Zoology, ASUMZ 13027 (juvenile female), collected by C.T. McAllister 890422–8, 22 April, 1989.

Entymology: The specific epithet is derived from *spina* (Latin for "thorn") to denote the spine-like, conical tubercles along the anterior edge of the carapace of the host turtle.

Remarks: Sporulated oocysts of this species are similar in size and shape to those of *E. carri* from the eastern box turtle, *Terrapene carolina* in Alabama and Florida (Ernst & Forrester, 1973) and *E. ornata* from ornate box turtles, *T. ornata* in Texas (McAllister & Upton, 1989). However, the sporulated oocysts are easily differentiated between these other species by having filaments on the SB end of their sporocysts. Other species of turtle coccidians known to have SB with filaments are *E. filamentifera* from snapping turtles (*C. serpentina*), *E. trachemydis* from red eared sliders (*T. scripta elegans*), *E. caretta* from loggerhead sea turtles (*C. caretta*), and *E. amydae* and *E. pallidus* also from *A. s. pallidus*. The sporulated oocysts of this species do not resemble any of those from the six eimeriid species previously reported from other softshell turtles of the Old and New Worlds (Roudabush, 1937; Chakravarty & Kar, 1943; Wacha & Christiansen, 1976, 1977).

Eimeria vesticostieda Wacha & Christiansen, 1977

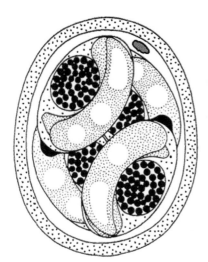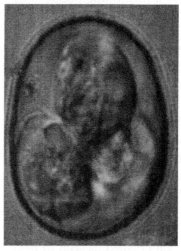

Figures 2.99, 2.100 Line drawing of the sporulated oocyst of Eimeria vesticostieda *(original) adapted from Wacha and Christiansen (1977). Photomicrograph of a sporulated oocyst of* E. vesticostieda *from Wacha and Christiansen (1977), with permission from John Wiley & Sons, Ltd. holder of the copyright for the* Journal of Eukaryotic Microbiology *(formerly* Journal of Protozoology*).*

Type host: *Apalone spinifera* (La Sueur, 1827) (syn. *Trionyx spiniferus* Le Sueur), Eastern Spiny Softshell Turtle.

Type locality: NORTH AMERICA: USA: Iowa, Polk County, Chichaqua Wildlife Refuge, Secs. 28−29−33, T81N, R22W.

Other hosts: *Apalone spinifera hartwegi* (Conant & Goin, 1941), Western Spiny Softshell Turtle.

Geographic distribution: NORTH AMERICA: USA: Iowa.

Description of sporulated oocyst: Oocyst shape: narrowly ovoidal to narrowly ellipsoidal; number of walls: 1; wall characteristics: outer surface smooth, ~1.3 thick; L × W (N = 27): 23.4 × 18.6 (22−25.5 × 16.5−20.5); L/W ratio: 1.3 (1.1−1.4); M, OR: both absent; PG: present, but small (line drawing). Distinctive features of oocyst: sporocysts are tightly packed with the oocyst.

Description of sporocyst and sporozoites: Sporocyst shape: ellipsoidal to slightly ovoidal; L × W (N = 27): 13.6 × 8.1 (12−15.5 × 7.5−9); L/W ratio: 1.7; SB: a thick, convex, vesicle-like body; SSB, PSB: both absent; SR: present; SR characteristics: ellipsoid to spherical, 4.5−7.5 wide, a membrane-bounded body filled with spheroidal granules each ∼0.5; SZ: usually with two (sometimes three) spheroidal or ellipsoidal RBs, but the N is not discernible. Distinctive features of sporocyst: thick, convex, vesicle-like SB and a membrane-bounded SR.

Prevalence: Unknown. Not stated in original description.

Sporulation: Unknown.

Prepatent and patent periods: Unknown.

Site of infection: Unknown. Oocysts found in feces.

Endogenous stages: Unknown.

Cross-transmission: None to date.

Pathology: Unknown.

Materials deposited: None.

Remarks: Only *E. paynei* has sporulated oocysts which, like those of *E. vesicostieda*, are narrowly rounded, of similar size, and have a PG, an SB, and a wall of about the same thickness, which lacks an M. However, the sporulated oocysts of *E. vesicostieda* differ from those of *E. paynei* by having a wall of uniform thickness which, when crushed, is seen to have one rather than two layers, the shapes are somewhat different, and it has SZ with more than one RB. In addition, the hosts of the two species live in contrasting ecologic habitats, so there seems little opportunity in nature, for cross-transmission of oocysts between these two host species. The members of the former are aquatic, occupying primarily rivers, lakes, and ponds, while those of the latter are terrestrial, occupying high ground comprised typically of sandy ridges and dunes in areas where the water table does not normally come to the surface.

Genus *Chitra* Gray, 1831 (3 Species)

To our knowledge, there are no coccidia described from this genus.

Genus *Cyclanorbis* Gray, 1854 (2 Species)

To our knowledge, there are no coccidia described from this genus.

Genus *Cycloderma* Peters, 1854 (2 Species)

To our knowledge, there are no coccidia described from this genus.

Genus *Dogania* Gray, 1844 (Monospecific)

To our knowledge, there are no coccidia described from this genus.

Genus *Lissemys* (Gray, 1831) Smith, 1931 (2 Species)
Eimeria innominata Kar, 1944

Figure 2.101 Line drawing of the sporulated oocyst of Eimeria innominata *(original) modified from Mandal (1976).*

Type host: *Lissemys punctata* (Bonnaterre, 1789) (syn. *Emyda granosa* Boulenger, 1899), Indian Flap-Shelled Turtle.

Type locality: ASIA: India: Calcutta.

Other hosts: None to date.

Geographic distribution: ASIA: India.

Description of sporulated oocyst: Oocyst shape: subspheroidal; number of walls: 2; wall characteristics: double contoured, the outer being more prominent than the inner; L × W: 17.7 × 13.5 (16.5–19 × 11.5–14); L/W ratio: 1.3; M, OR, PG: all absent (line drawings of Kar, 1944; Mandal, 1976). Distinctive features of oocyst: thin, double-layered wall.

Description of sporocyst and sporozoites: Sporocyst shape: pyriform with an irregular-shaped knob at the pointed end; L × W: 11.3 × 6.5 (10.5–12 × 5.5–7.5); L/W ratio: 1.7; SB: present at pointed end, but of "irregular-shape;" SSB; possibly present (line drawing, Mandal, 1976); PSB: absent; SR: present; SR characteristics: a little bunch of refringent globules; SZ: elongated, 6.6 × 4.4 (6.5–7.5 × 4–5), with one end narrowing to a sharp point while the other end is round; N: spheroidal, near center of SZ or closer to rounded end. Distinctive feature of sporocysts: the irregular knob shape of the SB/SSB (?) complex.

Prevalence: Unknown.

Sporulation: Unknown.
Prepatent and patent periods: Unknown.
Site of infection: Liver (Mandal, 1976).
Endogenous stages: Unknown.
Cross-transmission: None to date.
Pathology: Unknown.
Materials deposited: None.
Remarks: Measurements used in the above descriptions are taken mostly from Mandal (1976), but Kar's (1944) are used where needed to supplement. The distinctive SB and SSB (?) complex at the tip of each sporocyst seems to distinguish this eimerian from others found in turtles. Sporulated oocysts of this species somewhat resemble those of *E. dericksoni* and *E. legeri* in shape and size. However, they differ from those of *E. dericksoni* in the absence of an OR, shape of the sporocyst, and the structure of the SZ. They differ in the shape and size of the sporocysts of *E. legeri*.

Eimeria irregularis Kar, 1944

Figure 2.102 Line drawing of the sporulated oocyst of Eimeria irregularis *(original) modified from Mandal (1976).*

Type host: *Lissemys punctata* (Bonnaterre, 1789) (syn. *Emyda granosa* Boulenger, 1899), Indian Flap-Shelled Turtle.
Type locality: ASIA: India: Calcutta.
Other hosts: None to date.
Geographic distribution: ASIA: India.
Description of sporulated oocyst: Oocyst shape: spheroidal to irregular; number of walls: 2; wall characteristics: extremely thin and transparent; L × W: 15.4–15.5 (15–16.5); L/W ratio: 1.0; M, OR, PG: all absent. Distinctive features of oocyst: according to Kar (1944), the irregular shape of the mature oocysts is a peculiar diagnostic character.

Description of sporocyst and sporozoites: Sporocyst shape: elongate-ovoidal with rounded posterior and somewhat pointed anterior end; also described as spindle-shaped (Kar, 1944); L × W: 12.5 × 6.9 (11.5–13.5 × 6.5–7.5); L/W ratio: 1.8; SB: present at pointed end of sporocyst; SSB, PSB: both absent; SR: present; SR characteristics: a compact mass of granules in center of sporocysts between SZ (line drawings of both Kar, 1944; Mandal, 1976); SZ: sausage-shaped with both the ends rounded, ~8.5–8.8 × 2.2–2.5 (Kar, 1944; Mandal, 1976); N: spheroidal, in center of SZ. Distinctive features of sporocysts: none, other than its spindle-shape.

Prevalence: Unknown.

Sporulation: 30–45 h (Mandal, 1976) or 48 h (Kar, 1944).

Prepatent and patent periods: Unknown.

Site of infection: Intestine (Mandal, 1976).

Endogenous stages: Unknown.

Cross-transmission: None to date.

Pathology: Unknown.

Materials deposited: None.

Remarks: The sporulated oocysts of this species resemble those of *E. trionyxae* and *E. koormae* in general appearance and size, but they differ from the former in not having an OR. The absence of an M and the shape of both the sporocysts and SZ distinguish oocysts of this species from those of *E. koormae*.

Eimeria koormae Das Gupta, 1938a, b

Figure 2.103 Line drawing of the sporulated oocyst of Eimeria koormae *(original) modified from Mandal (1976).*

Type host: *Lissemys punctata* (Bonnaterre, 1789) (syn. *Emyda granosa* Boulenger, 1899), Indian Flap-Shelled Turtle.

Type locality: ASIA: Bangladesh: Jessore.

Other hosts: None to date.

Geographic distribution: ASIA: India.

Description of sporulated oocyst: Oocyst shape: spheroidal with thick walls; number of walls: 2; wall characteristics: inner is thicker than outer; L × W: 14.6 (13.5−16); L/W ratio: 1.0; M: present; OR, PG: both absent; Distinctive features of oocyst: very thick bilayered wall with the inner layer being the thickest.

Description of sporocyst and sporozoites: Sporocyst shape: spindle-shaped, distinctly tapered at both ends; L × W: 10.3 × 4.6 (9−11 × 3.5−6); L/W ratio: 2.2; SB, SSB, PSB: all apparently absent; SR: present; SR characteristics: a granular mass situated between the SZ; SZ: with one rounded end that is stouter than the other. Distinctive features of sporocysts: distinctly-shaped spindles.

Prevalence: Unknown.

Sporulation: Exogenous; oocysts kept in 1% chromic acid for 2−3 days became completely sporulated (Das Gupta, 1938; Mandal, 1976).

Prepatent and patent periods: Unknown.

Site of infection: Das Gupta (1938) said that this species was totally confined to the cells of the intestine and that merogony took place in the epithelium, while all other endogenous stages occurred in the subepithelium.

Endogenous stages: Das Gupta (1938) fixed pieces of intestine and sectioned them to determine whether endogenous stages might be present. Early meronts were round, 2 × 2 wide, while fully mature meronts were only 4.1 wide and contained eight merozoites, each measuring 1.5 × 1.0. Das Gupta (1938) could not differentiate between early microgamonts and meronts, but when they had grown larger, the N of the microgametocyte divided into a large number of smaller N, which then gave rise to many microgametes clustered around a residual mass of cytoplasm. Macrogametocytes were spheroidal and contained darkly stained granules in their cyto-plasm even at their earliest stage. When fully formed, they were 12.4 wide. Completely formed, but unsporulated oocysts were sphe-roidal, ∼14 wide in the feces and rectal contents.

Cross-transmission: None to date.

Pathology: Unknown.

Materials deposited: None.

Remarks: Das Gupta (1938a) first named this species and partially characterized it in an abstract presented in the Proceedings of the Indian Science Congress, making it a *species inquirenda*; however, later that year (Das Gupta, 1938b) it was published with line drawings to establish it as a valid species. This species differs from other eimerians from turtles by the thick inner wall of its sporulated oocysts and their spindle-shaped sporocysts.

Eimeria légeri (Simond, 1901) Reichenow, 1921

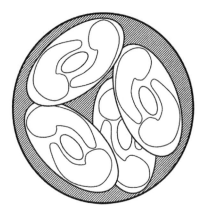

Figure 2.104 Line drawing of the sporulated oocyst of Eimeria légeri *(original) modified from Simond (1901).*

Synonym: *Coccidium légeri* Simond, 1901.

Homonym: *Eimeria légeri Stankovitch*, 1920 (syn. *Eimeria stankovitchi*).

Type host: *Lissemys punctata* (Bonnaterre, 1789) (syns. *Cryptopus granosus* Duméril & Bibron, 1835; *Emyda granosa* Boulenger, 1899), Indian Flap-Shelled Turtle.

Type locality: ASIA: India (presumably), but not stated in original description.

Other hosts: None to date.

Geographic distribution: ASIA: India (?).

Description of sporulated oocyst: Oocyst shape: spheroidal to slightly subspheroidal; number of walls: 1; wall characteristics: thin; L × W: 16–18 (Simond, 1901); L/W ratio: not stated, but probably ~1.0+; M, OR, PG: all absent in original description and line drawing (Simond, 1901). Distinctive features of oocyst: sporocysts are tightly packed into the confines of the oocyst wall, which is not distorted by their presence (Simond, 1901).

Description of sporocyst and sporozoites: Sporocyst shape: ovoidal; L × W: unknown, but extrapolating from the original drawing they are ~8–9 × 4–5; L/W ratio: unknown; SB: absent in original line drawing (Simond, 1901); SSB, PSB: both absent; SR: absent in original description, but the line drawing (Simond, 1901) shows a small, ellipsoidal body near the side of each sporocyst; SR characteristics: granular in the beginning and later becomes transparent and refringent; SZ: comma-shaped.

Prevalence: In 1/1 (100%) of the type host (Simond, 1901).

Sporulation: Unknown.

Prepatent and patent periods: Unknown.

Site of infection: Gall bladder, bile duct (Mandal, 1976), and/or liver (Pellérdy, 1974).

Endogenous stages: Simond (1901) said he saw young macrogametes in liver cells that were 16−18 wide; he was likely looking at a fresh tissue squash and he provided no line drawing or photomicrograph to substantiate his finding.

Cross-transmission: None to date.

Pathology: Unknown.

Materials deposited: None.

Remarks: The original description by Simond (1901) was incomplete with no mensural data and was virtually useless except that he provided a crude line drawing of a sporulated oocyst, a sporocyst, and a sporozoite (his Figures 4−6). Mandal (1976) repeated another totally incomplete description and said that he examined 15 hosts from different localities, but none of them were infected; thus, he based his few descriptive statements on Bhatia (1938), who also wrote virtually nothing to further characterize this species.

Genus *Nilssonia* Gray, 1831 (5 Species)

Eimeria triangularis Chakravarty & Kar, 1943

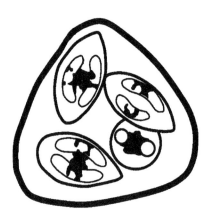

Figure 2.105 Line drawing of the sporulated oocyst of Eimeria triangularis *(original) modified from Mandal (1976).*

Type host: *Nilssonia gangetica* (Couvier, 1825) (syn. *Trionyx gangeticus* Cuvier, 1825), Ganges Softshelled Turtle.

Type locality: ASIA: India: Calcutta, purchased from a market.

Other hosts: None to date.

Geographic distribution: ASIA: India.

Description of sporulated oocyst: Oocyst shape: usually triangular, but a few appeared somewhat spindle-shaped; number of walls: 1 (original line drawing); wall characteristics: outer surface is smooth and apparently flexible; L × W: 10.3−14.4 in longest diameter (Chakravarty & Kar, 1943) or 12.5 (9.5−14.5) (Mandel, 1976); L/W ratio: 1.0 (?); M, OR, PG: all absent. Distinctive features of oocyst: triangular shape and small size.

Description of sporocyst and sporozoites: Sporocyst shape: ovoidal (early) to spindle-shaped (mature), with both ends bluntly rounded; L × W: 10.3 × 4.1 (Chakravarty & Kar, 1943) or 10.5 × 3.1 (9.5−11.5 × 3−5); L/W ratio: 3.0 (Mandal, 1976); SB, SSB, PSB: all absent; SR: present; SR characteristics: a granular mass between SZ; SZ: elongated, ∼9.5 long. Distinctive features of sporocyst: spindle-shape and small size.

Prevalence: In 1/2 (50%) of the type host (Chakravarty & Kar, 1943).

Sporulation: 24−36 h (Mandal, 1976).

Prepatent and patent periods: Unknown.

Site of infection: Intestine (Mandal, 1976).

Endogenous stages: Unknown.

Cross-transmission: None to date.

Pathology: Unknown.

Materials deposited: None.

Remarks: The triangular shape of these oocysts along with the spindle-shape of their sporocysts make the sporulated oocyst of this coccidium unique among all others, if it exists. Oocysts resembling this form have not been reported again since its original description >70 years ago.

Eimeria trionyxae Chakravarty & Kar, 1943

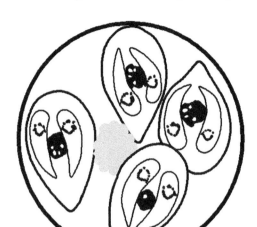

Figure 2.106 Line drawing of the sporulated oocyst of Eimeria trionyxae *(original) modified from Mandal (1976).*

Type host: *Nilssonia gangetica* (Couvier, 1825) (syn. *Trionyx gangeticus* Cuvier, 1825), Ganges Softshelled Turtle.
Type locality: ASIA: India: Calcutta, purchased from a market.
Other hosts: None to date.
Geographic distribution: ASIA: India.
Description of sporulated oocyst: Oocyst shape: spheroidal; number of walls: one; wall characteristics: very thin (line drawing, Mandal, 1976); L × W: 16.5 (14–19); L/W ratio: 1.0; M, PG: both absent; OR: present, as noted in Table I of Chakravarty and Kar (1943), and in their original "diagnosis," but it is absent in their original line drawing (their Figure 4); Mandal (1976) also said it is present, but his line drawing also shows it to be absent. Distinctive features of oocyst: spheroidal shape, a very thin, single-layered wall (line drawings of Chakravarty & Kar, 1943; Mandal, 1976), and presence of an OR (not shown in their line drawings).
Description of sporocyst and sporozoites: Sporocyst shape: pyriform to tear-drop shaped, with one end rounded while the other is pointed; L × W: 12.4 × 6.2 (11.5–13 × 6–8); L/W ratio: 2.0; SB, SSB, PSB: all reported as absent, but it is possible that a small SB may be found at the pointed end of sporocyst; SR: present; SR characteristics: a centrally placed spheroidal mass of small granules

between the SZ: SZ: narrow, elongated, ∼9.5–10.5 long, with a central N. Distinctive features of sporocysts: tear-drop shaped with a spheroidal SR centrally placed between the SZs.

Prevalence: In 2/2 (100%) of the type host (Chakravarty & Kar, 1943).

Sporulation: Exogenous. 48–60 h (Mandal, 1976).

Prepatent and patent periods: Unknown.

Site of infection: Unsporulated oocysts were found in large numbers in the rectum of the type host (Chakravarty & Kar, 1943), but both Chakravarty & Kar (1943) and Mandal (1976) said the developmental stages were scattered throughout the epithelial cells of the intestine.

Endogenous stages: Although their methods weren't stated, Chakravarty and Kar (1943) apparently fixed and embedded intestinal tissue for histological examination. They reported trophozoites, 7.4 × 4.1, as ovoid bodies with one end rounded and the other tapering (their Figure 5). The tapering end was directed toward the submucosa; it had an N in its midbody. Meronts were ovoidal to spheroidal, 8.2–14.4 wide. Arrangement of the merozoites within the meronts was constant in all observations with their more rounded ends converging in the center of the meront, while the slightly more pointed end was directed toward its periphery, which they compared to the rosetta formation. There was a residual body in the center of the mature meront. Fully formed merozoites were elongated and club-like with one end tapering slightly; each had a rectangular N. Merozoites were 4.1–6.2 × 1. Mature microgamonts, although similar in size to meronts, were easily distinguished from them by the structure of their N; these were smaller in size and vesicular. As the N elongated they became thread-like bodies and when mature, each had a long flagellum attached at one end. Macrogamonts were spheroidal with granulated cytoplasm, a circular N, and a large eccentric nucleolus.

Mandal (1976) said that he collected and examined the topotype host and also found some endogenous stages of this eimerian, but the description and measurements presented were identical to those given by Chakravarty and Kar (1943). Mandal (1976) said (i.e., repeated) that the trophozoites (presumably, early meronts just after entering a host cell) were ovoidal bodies, 7.4 × 4.1, with one end tapering, the other rounded, and a spheroidal N in the middle. Mature meronts were 8.2−14.4 wide, spheroidal to ovoidal with "a large number of nuclei of uniform size." Merozoites were elongate, club-shaped bodies, 4.1−6.2 × 1.0, and were arranged in the form of a rosette. Microgametes were rod-like bodies with a long flagellum at one end. Macrogametes were spheroidal with a like-shaped N enclosed by a nuclear membrane, and had a small, central nucleolus (Mandall, 1976).

Cross-transmission: None to date.
Pathology: Unknown.
Materials deposited: None.
Remarks: The size of the oocysts and sporocysts as well as the pyriform to tear-drop shaped sporocysts makes the structural features of this species unique.

Genus *Palea* Meylan, 1987 (Monospecific)
To our knowledge, there are no coccidia described from this genus.

Genus *Pelochelys* Gray, 1864 (3 Species)
To our knowledge, there are no coccidia described from this genus.

Genus *Pelodiscus* Fitzinger, 1835 (4 Species)
To our knowledge, there are no coccidia described from this genus.

Genus *Rafetus* Daudin, 1802 (2 Species)
To our knowledge, there are no coccidia described from this genus.

Genus *Trionyx* Saint-Hilaire, 1809 (Monospecific)
To our knowledge, there are no coccidia described from this genus.

SUPERFAMILY KINOSTERNOIDEA

FAMILY DERMATEMYDIDAE, RIVER TURTLES, 1 GENUS, MONOTYPIC

Genus *Dermatemys* Gray, 1847 (Monospecific)
To our knowledge, there are no coccidia described from this genus.

FAMILY KINOSTERNIDAE, MUD & MUSK TURTLES, 4 GENERA, 25 SPECIES

Genus *Claudius* Cope, 1865 (Monospecific)

To our knowledge, there are no coccidia described from this genus.

Genus *Kinosternon* Spix, 1824 (18 Species)

Eimeria lutotestudinis Wacha & Christiansen, 1976

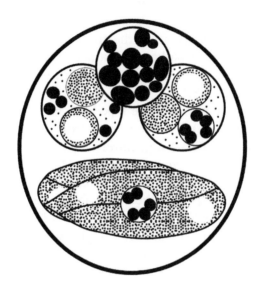

Figure 2.107 Line drawing of the sporulated oocyst of Eimeria lutotestudinis *(original) modified from Wacha and Christiansen (1976).*

Type host: *Kinosternon flavescens spooneri* Wermuth & Mertens, 1977, Yellow Mud Turtle (also, Illinois Mud Turtle).

Type locality: NORTH AMERICA: USA: Iowa, Smith.

Other hosts: *Graptemys caglei* Haynes & McKnown, 1974, Cagle's Map Turtle; *Graptemys geographica* (Le Sueur, 1817), Common Map Turtle; *Kinosternon flavescens flavescens* Warmuth & Mertens, 1977, Yellow Mud Turtle; *Kinosternon subrubrum hippocrepis* Gray, 1855, Mississippi Mud Turtle; *Pseudemys texana* Baur, 1893, Texas River Cooter; *Trachemys scripta elegans* (Wied, 1838), Red Eared Slider.

Geographic distribution: NORTH AMERICA: USA: Iowa, Texas.

Description of sporulated oocyst: Oocyst shape: broadly ellipsoidal to subspheroidal, occasionally spheroidal; number of walls: 1; wall characteristics: ~0.5 thick, with smooth outer surface;

L × W: 11.9 × 10.8 (10–13 × 9–12); L/W ratio: 1.1 (1.0–1.2); M: absent; OR: spheroidal or ellipsoidal, membrane-bound, 2.5–5.0, with spheroidal or ellipsoidal granules, each ∼1.0; PG: absent. Distinctive features of oocyst: containing a membrane-bound OR.

Description of sporocyst and sporozoites: Sporocyst shape: narrowly ellipsoidal; L × W: 7.9 × 3.7 (7–10 × 3–4); L/W ratio: 2.6; SB: small, conical; SSB, PSB: both absent; SR: present; SR characteristics: spheroidal membrane-bound, 2.0–2.5, with a few loosely packed spheroidal granules, 0.3–0.5, that may be unbound and scattered within the sporocyst; SZ: with one spheroidal RB, ∼2.0, at broad end. Distinctive features of sporocyst: ellipsoidal shape with large L/W ratio and presence of membrane-bounded SR.

Prevalence: In 4/26 (15%) of the type host; in 1/7 (14%) *G. geographica* and in 1/6 (17%) *K. s. hippocrepis* from Arkansas; and in 5/13 (38%) *G. caglei*, 1/5 (20%) *K. f. flavescens*, 1/9 (11%) *P. texana*, and in 31/100 (31%) *T. s. elegans* from Texas (McAllister et al., 1994).

Sporulation: Likely exogenous, but unknown because Wacha and Christiansen (1976) placed the turtle feces into specimen jars in 2.5% aqueous (w/v) $K_2Cr_2O_7$ solution at ∼22°C for 1 week and then examined them for oocysts. Samples with sporulated oocysts were then stored in a refrigerator at ∼5°C.

Prepatent and patent periods: Unknown.

Site of infection: Unknown.

Endogenous stages: Unknown.

Cross-transmission: None to date.

Pathology: Unknown.

Materials deposited: None.

Entymology: The specific epithet of this eimerian is of Greek derivation to mean "of the mud turtle."

Remarks: Several *Eimeria* species reported from turtles have sporulated oocysts which, like *E. lutotestudinis*, have an OR, are broadly ellipsoidal to subspheroidal, lack an M, lack conical projections of the oocyst wall, and have ellipsoidal sporocysts. Those from *E. lutotestudinis* differ from *E. carri* by having smaller oocysts (11.9 × 10.8 vs. 15.9 × 14.5) and shorter sporocysts (7.9 vs. 11.1); from *E. dericksoni* by having smaller oocysts (11.9 × 10.8 vs. 14.6 × 12.9), by lacking a PG, and by having an SB, which appears more conical and pointed; and from those of *E. megalostiedai* by having more narrowly ellipsoidal sporocysts and an SB that is smaller.

Genus *Staurotypus* Wagler, 1830 (2 Species)

To our knowledge, there are no coccidia described from this genus.

Genus *Sternotherus* Gray, 1825 (4 Species)

To our knowledge, there are no coccidia described from this genus.

SUPERFAMILY CHELONIOIDEA

FAMILY CHELONIIDAE, SEA TURTLES, 5 GENERA, 6 SPECIES

Genus *Caretta* Rafinesque, 1814 (Monospecific)
Eimeria caretta Upton, Odell, & Walsh, 1990

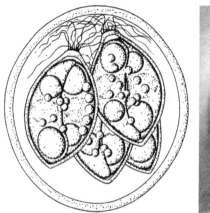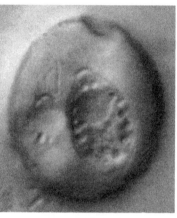

Figures 2.108, 2.109 Line drawing of the sporulated oocyst of Eimeria caretta *from Upton et al. (1990), with permission from the* Canadian Journal of Zoology. *Photomicrograph of a sporulated oocyst of* E. caretta *from Upton et al. (1990), with permission from the* Canadian Journal of Zoology.

Type host: *Caretta caretta* (L., 1758), Loggerhead Turtle.
Type locality: NORTH AMERICA: USA: Florida, Martin County, Huchinson Island, marine water surrounding Jensen Beach.
Other hosts: None reported to date.
Geographic distribution: NORTH AMERICA: USA: Florida.
Description of sporulated oocyst: Oocyst shape: subspheroidal, rarely ellipsoidal; number of walls: 2; wall characteristics: smooth, colorless, outer wall thin ($\sim$0.4) while inner wall is thicker ($\sim$0.6); L × W: 24.5 × 22.0 (21−28 × 18−24); L/W ratio: 1.1 (1.0−1.3); M, OR, PG: all absent. Distinctive features of oocyst: mostly spheroidal shape.

Description of sporocyst and sporozoites: Sporocyst shape: ovoidal; L × W: 14.3 × 8.9 (13−16 × 8−10); L/W ratio: 1.6 (1.5−1.8); SB: ~0.8 high × 3−3.5 wide, with 12−20 long, thin filaments that arise from the surface; SSB, PSB: both absent; SR: present; SR characteristics: granules scattered among SZ; SZ: elongate, 11.3 × 3.8 (10−12 × 3−4) *in situ*, arranged head-to-tail within sporocyst, with spheroidal to subspheroidal anterior RB, 2.6 × 2.5 (2−3 × 2−3) and a larger, spheroidal to ellipsoidal posterior RB, 3.8 × 3.2 (3−4 × 2−4); N indistinct, situated between RBs. Distinctive features of sporocyst: large sporocysts with 12−20 long, thin filaments arising from the SB of each.

Prevalence: In 2/2 (100%) of the type host.

Sporulation: Probably exogenous, but unknown since oocysts were placed into 2.5% aqueous (w/v) $K_2Cr_2O_7$ solution and not observed until 3 days later; after 3 days, some oocysts remained unsporulated, some were partially sporulated, and some were completely sporulated.

Prepatent and patent periods: Unknown.

Site of infection: Unknown. Oocysts recovered from feces.

Endogenous stages: Unknown.

Cross-transmission: None to date.

Pathology: No mention was made of pathology because the authors had access only to fecal samples.

Materials deposited: Syntypes (oocysts in 10% formalin) are deposited in the USNPC, Beltsville, Maryland, USA, No. 80927.

Entymology: The specific epithet reflects that of the host genus.

Remarks: The filaments found associated with the SB are rare, but not unique. Similar structures have been reported for *Eimeria cyclopion* McAllister, Upton, & Trauth, 1990, from *Nerodia cyclopion cyclopion* (a green water snake, see Duszynski & Upton, 2009), *E. trachemydis* from the red eared slider, *T. s. elegans*, and *E. filamentifera* from the common snapping turtle, *C. s. serpentina*. The former two species have oocysts and sporocysts that are more elongate than those in this species, L/W 1.4 (1.1−1.6) and 1.8 (1.5−2.2), respectively, vs. 1.1 (1.0−1.3) in *E. caretta* (see McAllister & Upton, 1988; McAllister et al., 1990). *Eimeria caretta* is distinguished from *E. filamentifera* because the former has more numerous and elongate filaments on the SB, lacks the bottleneck-like appearance of the anterior portion of the sporocyst, has a diffuse rather than compact SR, and does not have an OR.

Genus *Chelonia* Brongniart, 1800 (Monospecific)
Caryospora cheloniae Leibovitz, Rebell, & Boucher, 1978

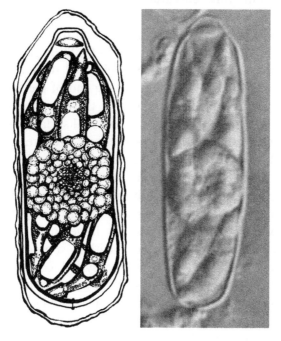

Figures 2.110, 2.111 Line drawing of the sporulated oocyst of Caryospora cheloniae *from Leibovitz et al. (1978), with permission from the* Journal of Wildlife Diseases. *Photomicrograph of a sporulated oocyst of* C. cheloniae *from Leibovitz et al. (1978), with permission from the* Journal of Wildlife Diseases.

Synonym: *Caryospora* sp. of Robell, 1974.

Type host: *Chelonia mydas mydas* (L. 1758), Green Turtle.

Type locality: CARIBBEAN: British West Indies: Grand Cayman.

Other hosts: None to date.

Geographic distribution: AUSTRALIA: Queensland; CARIBBEAN: British West Indies, Grand Cayman.

Description of sporulated oocyst: Oocyst shape: elongate-ellipsoidal with long axis often curved to varying degrees, like a sausage or cucumber; number of walls: 2; wall characteristics: thin 0.3−0.6 (0.5), smooth, very fragile and irregular, and during sporulation it usually disintegrates, releasing the sporulated sporocyst; only 6/1,000 sporulated sporocysts were found contained within their oocyst wall (Kinne et al., 1978); L × W: 37.4 × 12.8 (34−40 × 11−15); L/W ratio: 2.9; M, OR, and PG: all absent. Distinctive features of oocyst: fragile, 2-layered wall that disintegrates during sporulation releasing sporulated sporocysts.

Description of sporocyst and sporozoites: Sporocyst shape: elongate with long axis often curved to varying degrees, like a sausage or cucumber and this curvature alters the L/W ratio; L × W: 34.5 × 12.7 (26–44 × 11–17); L/W ratio: 2.8 (1.5–4.2); SB: distinctive, with a thin layer of the sporocyst wall that formed an elevated dome, 2.0 high × 4.6 wide (1–3.5 × 2–6); SSB: present as a colorless, lenticular body under the SB (elevated dome), 1.3 high × 3.8 wide (1–2 × 2–5); SSB: absent, but the end of the sporocyst opposite SB with a fine, median fissure that traversed the sporocyst wall perpendicularly from its internal to external surface with ends of fissure elevated as tiny knob-like projections; SR: present; SR characteristics: 11.4 × 9.7 (9–17 × 7–12), spheroidal, composed of coarse granules at its surface and less dense, finer granules internally, and is usually central in location with the SZ extending toward poles of sporocyst; upon storage, SR migrates toward one pole with SZ displaced toward other end, while the large outer granules of SR become detached, floating freely within sporocyst; SZ *in situ* (N = 25): 13.3 × 2.8 (13–17 × 2–4); excysted, completely extended, SZ (N = 2) were 15.8 × 3.8 (15–16 × 3.8), club or cigar-shaped with a blunt, rounded end and an apical pointed end capable of elongation and flexion; two RBs: posterior, 2.9 × 2.5 (2–4 × 1–3) as a prominent elongated elevated structure with a ring-like structure behind it while anterior RB was ~1.6 (1–2 wide); N: small, round, poorly defined, between RBs. Distinctive feature of sporocyst: SR migrates toward one pole displacing SZs toward the other end and the large outer granules of SR become detached, floating freely within sporocyst.

Prevalence: Not stated in original description as oocysts were harvested from feces and necrotic intestinal mucosa of stock hatchlings and juvenile tank-reared green sea turtles (Leibovitz et al., 1978); in 24/24 (100%) of *C. mydas* selected from 70 wild green turtles that had died in Moreton Bay, southeast of Queensland over a 6 week period in the spring, 1991 (Gordon et al., 1993a, b); three further cases of infection in sea turtles were found during October–December, 1992, indicating a seasonality of occurrence according to Gordon et al. (1993a, b).

Sporulation: Some oocysts (<1%) were completely sporulated in 19.5 h at 25°C when placed in 2.5% w/v $K_2Cr_2O_7$, but even after 98 h, only 92/672 (14%) oocysts had sporulated. After this time, little change was noted (Leibovitz et al., 1978). Gordon et al. (1993a, b) put oocysts in filtered seawater in Petri dishes for 5 days at 28°C, but said that sporulation was completed in 15–24 h. Within 24 h, the sporocyst split transversely into two unequal parts and all SZs "glided out." These excysted SZ were reported to arrange themselves into a star-like pattern, all attached by their obtuse ends, that looked something like a radiolarian. These star-like structures remained in this formation for 2–3 days, after which they disintegrated (Gordon et al., 1993b).

Prepatent and patent periods: Unknown.

Site of infection: Epithelial cells of the posterior third of the gastro-intestinal tract that includes the lower small intestine and the large intestine. Active patches of infection also are found in the upper (small) intestine of surviving and apparently healthy turtles as late as 1 month after the end of an epidemic and die off (Leibovitz et al., 1978). In Australia, Gordon et al. (1993a, b) found parasites in cells along the entire length of the intestine except for the proximal duodenum. They also reported endogenous stages of the parasite in extraintestinal sites including the thyroid glands (white foci up to 1 mm), kidneys (hemorrhage), trachea (airway foam), and brain (inflammation of the meninges).

Endogenous stages: "Earliest and smallest stages seen" (time not given; see Leibovitz et al., 1978) in infected epithelium were short, cigar-shaped meronts (N = 10), 10.3×8.4 ($9-11 \times 8-9$), with one rounded and one pointed end; rounded end contained an N, 2.9 (2–4) wide, with a prominent nucleolus. These meronts were found in loosely arranged groups of 2–6 "in the middle of the epithelial layer" (Leibovitz et al., 1978). Binary fission of this stage resulted in transverse division of the N, nucleolus, and cytoplasm followed by elongation and separation of the two cells perpendicular to their long axis, while the segments remained attached, end-to-end following division, forming cords of divided cells. One end of each cord was closer to the intestinal lumen, while segments of the cord closer to the basement membrane were larger and could be differentiated as either immature microgamonts or macrogamonts.

Young macrogamonts had thin walls, a fine, eosinophilic granular cytoplasmic reticulum, and a small, round, central eosinophilic N containing a prominent basophilic nucleolus. Mature macrogamonts were ellipsoidal, 32.2×12.2 ($29-37 \times 10-14$) with an N, ~ 4.5 (4–6) wide, and a nucleolus, ~ 1.8 (1–2) wide. They had less dense cytoplasm and a vacuole around each, isolating it from the surrounding tissue. Mature macrogamonts were reported to be closer to the luminal surface (Leibovitz et al., 1978).

Young microgamonts were darkly stained, ovoidal to spheroidal structures. As they grew, the host cell became eccentrically displaced and elongated, compressed by the growing parasite. Continued development produced dark-staining, basophilic granules at the periphery and a central, dark-staining spheroidal body; mature microgamonts were 37.6×23.9 ($31-43 \times 16-27$). When they ruptured at the cell's luminal surface, they released myriads of biflagellated, curved rod-like microgametes <0.25 long.

Following fertilization, unsporulated oocysts were surrounded by a large clear space formed during migration from the ruptured host cell into the gut lumen.

Cross-transmission: None to date.

Pathology: Pathology is most pronounced in the posterior third of the intestine, that includes the posterior small and large intestines. The hindgut lumen became greatly dilated as it filled with blood, oocysts, and tissue debris and its wall was thinner than normal and became thinnest where the epithelial folds had sloughed into the lumen. The tips of the folds of the hindgut became denuded of their epithelial cover and free blood escaped from the blood vessels of the tunica propria into the intestinal lumen. Remnants of the epithelium were markedly altered and epithelial hyperplasia was pronounced at the margins of the denuded areas. Densely packed, irregularly shaped epithelial cells with large N and scant cytoplasm were present in these areas. Numerous inflammatory cells, including eosinophils, were present in the inflamed mucosal layer. The anterior and the middle thirds of the intestinal tract showed little structural change, although individual developmental stages of the parasite were seen within epithelial cells at the tips and sides of the villi of the small intestine (Leibovitz, et al., 1978). Rebell (1974) reported plaque-like lesions of the mucosa and gelatinous sloughs of mucosa and oocysts throughout the length of the intestinal tract distal to the opening of the bile duct. Infected turtles were typically flat, weak, and emaciated. Other symptoms of the disease were lethargy, weakness, and depression. Tube-like casts of intestinal mucosa and impaction of the gut with caseous masses of oocysts occurred late in the disease, although bacterial flora remained essentially typ-ical of turtles at the Grand Cayman farm (Rebell, 1974). Gordon et al. (1993a, b) reported severe effusive enteritis while the intestinal mucosa varied in color from deep crimson to yellow or green where it had been replaced by a friable diphtheritic membrane (Gordon et al., 1993b); they (1993b) also documented inflammation in extraintestinal sites including the thyroid gland, kidney, and brain. Meningoencephalitis due to the presence of the parasite (e.g., mer-onts around blood vessels in the brain) resulted in neurological disturbance that included circling, head-tilting, and nystagmus (involuntary eye movement). Histological sections showed menin-geal thickening with inflammatory cells and fibroblasts, as well as perivascular cuffs and randomly distributed foci in inflammatory cells such as granulocytes and lymphocytes in the parenchyma.

Materials deposited: None.

Entymology: The specific epithet reflects the genus name of the type host.

Remarks: Rebell (1974) first identified this caryosporan when he reported on an epidemic of coccidial disease in two groups of recently hatched green turtles in the spring of 1973, but he did not name it. The disease, and its associated mortality, appeared in young turtles about 30 days after hatching and ran a 60-day course through the stock hatchlings. The turtles were raised in a series of fiber glass tanks at Mariculture, Ltd., Grand Caymen, British West Indies. In addition to being the first *Caryospora* reported from tur-tles, the life cycle is distinctive because of its division by transverse binary fission and the structure of the sporulated oocysts. *Caryospora cheloniae* produces marked tissue destruction in the hindgut, is the first, and still only, serious coccidial pathogen reported from turtles, and it is an economically important pathogen of mariculture-reared green sea turtles (Leibovitz, et al., 1978). Rebell (1974) noted that the epidemic caused by this parasite was unprecedented in the history of the Mariculture, Ltd. farm and that the coccidium may have been introduced in sand brought from Ascension Island or from Surinam. It also is possible the parasite naturally infected turtles at the farm, but remained undetected until management circumstances permitted it to reach epidemic propor-tions. It is likely that once more and more turtles became infected and large numbers of oocysts were constantly being released into the well-aerated water, a condition that favored sporulation, that conditions were ripe to produce the epidemic that resulted. Gordon et al. (1993b) found something not reported in the original description of the oocyst and sporocyst by Leibovitz et al. (1978); they (1993b) said that no sporocysts were observed within oocysts and the sporocysts they measured (N = 19) were larger than those in the original description, 44×13 ($37-51 \times 10-16$), although the ranges did overlap. The other difference between the two reports of infection in the British West Indies (Leibovitz et al., 1978) and that reported from Australia (Gordon et al., 1993a, b) is that the former occurred only in juvenile turtles during cultivation in laboratory tanks, while the latter was found only in free living adults. Development in extraintestinal sites is not unprecedented in some coccidia (e.g., Novilla et al., 1981) and may be relatively common in poikilotherms (Overstreet, 1981). In fact, a number of studies have demonstrated facultatively heteroxenous life cycles of other *Caryospora* species involving secondary hosts in which extraintestinal development occurs (Lindsay & Sundermann, 1989; Dubey et al., 1990; Douglas et al., 1991, 1992).

Genus *Eretmochelys* Linnaeus, 1766 (Monospecific)
To our knowledge, there are no coccidia described from this genus.

Genus *Lepidochelys* Fitzinger, 1843 (2 Species)
To our knowledge, there are no coccidia described from this genus.

Genus *Natator* McCulloch, 1908 (Monospecific)
To our knowledge, there are no coccidia described from this genus.

FAMILY DERMOCHELYIDAE, LEATHERBACK TURTLES, 1 GENUS, MONOTYPIC

Genus *Dermochelys* Blainville, 1816 (Monospecific)
To our knowledge, there are no coccidia described from this genus.

DISCUSSION AND SUMMARY

Unlike the smaller suborder, Pleurodira, members of suborder Cryptodira have necks that are able to be retracted into the animals' shells rather than folding sideways. Members of this suborder make up the majority of living tortoises and turtles on Earth. These animals evolved during the Jurassic period with the oldest fossil record dating to approximately 190 million years ago (MYA) (Gafney et al., 1987). Today, members placed in this suborder are highly speciose and inhabit a wide variety of habitats all over the world. Within the Cryptodira are 11 families containing 73 genera and 246 species. Seven of the 11 (64%) families have representatives that have been examined for coccidia. Within these seven families, only 26/73 (36%) of the genera and only 57/246 (23%) of the species (not including subspecies designations) have been examined for coccidia. From these 57 turtle species, five apicomplexan genera have been recovered: *Caryospora* (two species), *Cryptosporidium* spp. (number of species is unknown; found in seven turtle species in six genera), *Eimeria* (60 named species; five unnamed species in five turtle species in four genera), *Isospora* (three named species), *Sarcocystis* (one named species; eight unnamed species in seven turtle species in three genera). Even with all of this published information on the coccidia from turtles, we still know almost nothing about their biology and how they affect these hosts.

Coccidia and Coccidiosis in Cryptodira

All but one of these species reports either never mentions obvious signs of pathology in the host from which feces were collected or they state that the host showed no outward signs of pathology. In retrospect, this is due to the fact that many of the coccidians described and named were retrieved from fecal samples rather than from necropsies, where internal lesions demonstrating pathology would be more easily identified. From all the *Eimeria* and *Isospora* species described (from oocysts), we know only a little about the endogenous stages of only four eimerians (*E. koormae, E. légeri, E. mitraria, E. trionyxae*) and no pathology was mentioned in any of them. Only *Cary. cheloniae* from green sea turtles (*Chel. m. mydas*) was examined in enough detail histologically from tissue sections such that severe pathology, directly attributed to the parasite, could be demonstrated. The absence of obvious pathology in the vast majority of studies on turtle coccidia also could be due to the genus and species of coccidian that infects the host organism, i.e., 63/64 (98%) of the named coccidian species found to infect turtles in this suborder belong to the genus *Eimeria* (60) or *Isospora* (3), which most often present asymptomatically in natural infections of wild hosts, although diarrhea can be present in heavy infections where such infected hosts will pass bloody stools. From the small amount of evidence that is available for us to study, as summarized above, it appears that these turtles rarely host eimerians and isosporans that cause pathology.

Such is not the case with the only caryosporan described from turtles. Rebell (1974) was the first to report pathology associated with a turtle coccidian when he identified an odd-looking caryosporan as the causative agent of a coccidiosis epidemic among populations of young, mariculture-reared green sea turtle (*Chel. m. mydas*) hatchlings. The causative agent of this epidemic was later identified and named *Cary. cheloniae* (Leibovitz et al.,1978) when it was again found to infect a group of mariculture-reared green sea turtle hatchlings. Leibovitz et al. (1978) further described the pathology induced by infection with *C. cheloniae* in detail. The majority of the damage caused by this parasite was found in the lower gastrointestinal tract, specifically in the hindgut of the hosts. Morphological discrepancies of the sporulated oocysts were published by Gordon et al. (1993b) after finding *Cary. cheloniae* in adult green sea turtles from Australia. These discrepancies are likely attributed to both environmental and developmental differences among the hosts studied. Leibovitz et al. (1978) studied hatchlings reared in a small, captive environment, whereas Gordon et al. (1993b) observed coccidia from wild adult green sea turtles. The close quarters and environmental conditions of the hatchlings' environment increased the ability of the parasites to infect most hosts. On the other hand, the wild adult turtles contracted the parasites from a much less parasite-dense environment with conditions that were more variable and perhaps less preferential for coccidian development. The age of the turtles also may have played a role in the morphological differences between the oocysts of the parasite, though the environmental differences likely had a larger impact on these differences.

Greiner (2003) reviewed the literature on coccidiosis reported from reptiles and noted what we stated earlier, namely, that most species are not considered pathogens even though they (their endogenous stages during both merogony and gamogony) all kill host cells. Apparently, gut cell turnover in vertebrate hosts, including turtles, is such that cells that are destroyed by endogenous stages are replaced quickly enough for the host to remain (or at least look) healthy. However, as we can see from the information presented in this chapter, the species of turtle coccidians are so poorly studied and understood that we have no way to access their level of pathogenicity. The major reason for this vacuum in our knowledge is that, with the exception of work on the green sea turtle (*Chelm. mydas*) discussed above, all but four other studies looked in fecal material for oocysts, but did not dissect hosts or look at tissue sections of infected gut to examine or document pathology. This is an area of study that some young scholar in parasitology could devote a lifetime to work on (and hopefully someone will!).

Endogenous Development

Everything we know about endogenous developmental stages in the 60 *Eimeria* species in the turtles of this suborder is based on four species: *E. koormae*, *E. légeri*, *E. mitraria*, and *E. trionyxae*. Simond (1901), who first saw *E. légeri*, was the first person to document an endogenous stage in a turtle (*L. punctata*), but his observation was done by looking at a tissue squash of liver and he only observed what he said was a young macrogametocyte. A year later, Laveran and Mesnil (1902) found and named *E. (C.) mitraria* in *M. reevesii*, but they also "followed fresh preparations" briefly described a meront with 20 merozoites, and what they believed were macrogametocytes and microgametocytes. Working on *E. koormae* from *L. punctata*, Das Gupta (1938) fixed pieces of intestine and sectioned them to search for endogenous stages that might be present. He found only one, small merogonous stage with eight merozoites when mature, and gave a brief description of both microgametocytes and macrogametocytes. Chakravarty and Kar (1943) discovered *E. trionyxae* in *N. gangetica* and prepared intestinal tissue from an infected turtle for histological examination; they briefly described trophozoites, one merogonous stage with club-like merozoites, and mature microgamonts and macrogamonts. Later, Mandal (1976) said that he also found some endogenous stages of *E. trionyxae*, but the description and measurements he presented were identical to those given by Chakravarty and Kar (1943). This is the sum of our current knowledge on enodogenous development of eimerians and isosporans in turtles.

To emphasize how incomplete our knowledge is, one need only remember that endogenous development almost always has two and usually four or more (asexual) merogonous stages, before gamete production begins and oocysts begin to leave the host in the feces. The time between when a susceptible host ingests infective oocysts and when the first newly produced oocysts are found in the feces is called the prepatent period. This prepatent period may be as short as 3–4 days or as long as 7–10 days, or longer. The length of time that oocysts continue to be shed in the feces, until cessation, is the patent period, which can last for days or weeks, depending on the coccidian species, the number of endogenous stages completed, and the size of the infecting dose. So when a wild animal is taken from the field and oocysts are found in its stool, there is no way to determine which day of the patent period it is experiencing. When meronts and gamonts are both present in host cells at the same time, it is likely that merogony is finishing its last generation and gamogony is well on its way to completing the patent period. Thus, we do not know one complete life cycle for any coccidian from turtles.

Treatment and Prevention

Little can be done to prevent wild turtles and tortoises from ingesting occasional sporulated oocysts that may enter into their environmental proximity via the natural food chain. In reptile collections, however, it is imperative to have quarantine facilities to keep newly acquired individuals from contaminating other animals in the collections. Many of the procedures and techniques outlined to keep snakes coccidia-free (Duszynski & Upton, 2009) apply equally as well to turtles as do the recommended therapies and treatment regimens (see their Table 12, pp. 312–316).

Host Specificity in Cryptodira Coccidia

The majority of turtle species from which coccidia have been discovered and named are known to harbor only *Eimeria* species (see Tables 1 and 2). Unlike what has been observed among mammalian coccidia, turtle eimerians seem to have little host specificity among some of their species, while others may be highly species-specific. For example, *E. mitraria* has been purportedly identified from 14 turtle species/subspecies spanning 12 host genera in three different families, demonstrating a low degree of host specificity, if these identifications are correct. Other presumably non host-specific eimerians include four identified from four host genera: *E. graptemydis* reported from *Chrysemys* (1 sp.), *Graptemys* (3 spp.), *Kinosternon* (2 spp.), and *Trachemys* (1 sp.); *E. lutotestudinis* reported from *Graptemys* (2 spp.), *Kinosternon* (2 spp.), *Pseudemys* (1 sp.), and *Trachemys* (1 sp.); *E. marginata* reported from *Chrysemys* (1 sp.), *Graptemys* (2 spp.), *Pseudemys* (1 sp.) and *Trachemys* (2 spp.); *E. pseudemydis* reported from *Deirochelys*

(1 sp.), *Glyptemys* (1 sp.), *Pseudemys* (1 sp.), and *Trachemys* (3 spp.). There also are four eimerians identified from three host genera: *E. chrysemydis* reported from *Chrysemys* (1 sp.), *Graptemys* (2 spp.), and *Trachemys* (2 spp.); *E. pseudogeographia* reported from *Chrysemys* (1 sp.), *Graptemys* (3 spp.), and *Trachemys* (2 spp.); *E. tetradacrutata* reported from *Chrysemys* (1 sp.), *Graptemys* (1 sp.), and *Trachemys* (1 sp.); and *E. trachemydis* reported from *Chrysemys* (1 sp.), *Graptemys* (1 sp.), and *Trachemys* (2 spp.). There is also one eimerian identified from two host genera: *E. arakanensis* reported from *Heosemys* and *Curora* (1 sp. each). All of the remaining 50 *Eimeria* species identified and named from turtles in this suborder are now known to infect only the single turtle species from which they were originally described (Table 1), although this could change with future studies. The only conclusion we can draw from these limited data is that host specificity may be highly variable among cryptodiran turtles. Experimental cross-transmission studies have not been performed for any of the parasites infecting turtles except for Carini (1942), who described the failure of attempts to infect other turtles with *Eimeria jaboti* from a *Chel. denticulata* by feeding them feces that contained sporulated oocysts.

Prevalence of Eimerians in Cryptodira

The relationships between coccidia and their vertebrate hosts have been evolving for millions of years. These parasites are thought to have emerged approximately 824 MYA (Escalante & Ayala, 1995). Diversity among anapsids (a term referring to a clade consisting of both extinct and extant chelonians based on their lack of temporal openings in the skull, which sets them apart from all other groups of reptiles) peaked about 260 MYA with the emergence of the first cryptodira ~185 MYA (Gaffney et al., 1987). Given this long history of probable association, it is not surprising that multiple genera and numerous species of coccidia are found to infect virtually all turtles that have been examined for them. Interestingly, there exists an overwhelming prevalence of eimerians within this group. Of the 65 valid, named coccidian species known to infect cryptodirid turtles, 60 are *Eimeria* species, while the remaining five are comprised of a single species each in the genera *Caryospora* (*C. cheloniae*) and *Sarcocystis* (*S. kinosterni*) and three species of *Isospora* (*I. chelydrae, I. rodriguesae,* and *I. testudae*).

Given the data that we have assembled about the coccidia infecting this group of turtles, it is evident that there remains a great deal of biological information and many new coccidian species yet to be discovered. It is highly probable that the vast majority of turtle species, and likely all of them, which have not been examined yet for coccidians, will be found to harbor one or more of these parasites.

Archiving Biological Specimens for Future Study

We are optimistic that, in the future, parasitologists will always archive actual specimens of both coccidian parasites and their hosts in accredited museums from which other scientists can access them and, if possible, their DNA, either fixed or with sequences stored in GenBank. Oocysts should be stored in 70–100% ethanol (EtOH) and hosts should be prepared as proper study skins (mammals, birds) or immersed in 70% EtOH (amphibians, fishes, invertebrates, reptiles) so that future workers can use them to amplify DNA as our techniques and protocols improve. Other methods to use for coccidia are histological sections of infected tissues on slides or pieces of infected tissues fixed and embedded in paraffin or plastic to be sectioned in the future. Finally, photosyntypes (digitized photomicrographs) of sporulated oocysts should always be archived in an accredited museum so other scientists can have access to them. Usually we would recommend the US National Parasite Collection (USNPC), Beltsville, Maryland, and/ or the Manter Parasitology Laboratory, Lincoln, Nebraska. However, after 121 years, there was a recent decision to transfer the USNPC to the National Museum of Natural History at the Smithsonian Institution. We've been told that an official announcement will be circulated sometime in 2014, and that, for the moment, the USNPC will be closed for about 12–18 months starting on July 1, 2014. We've also been told that "arrangements are being made for continuity in deposition of new material and for continued accession of type and voucher specimens," but no specifics have been given at the time of this writing. For the immediate future, we recommend all type specimens of coccidian materials be deposited into the H.W. Manter Parasite Collection at the University of Nebraska, Lincoln.

Parasitologists who have worked on coccidians from turtles have done a reasonably good job of trying to archive the hosts and/or parasites they've worked with. Of the 64 valid coccidia species named in this chapter, either host or parasite "type" materials, or both, have been archived for 36 (56%) in a museum. The symbiotype host (Frey et al., 1992) alone has been archived for three species; the symbiotype host and oocysts in formalin have been archived for five species; the symbiotype host and photosyntype (Duszynski, 1999) images of sporulated oocysts have been archived for three species; oocysts in formalin have been archived for five species; and photosyntypes of sporulated oocysts have been archived for 18 species. This is a good start, but we can, and should, continue to archive all protist materials to improve specimen collections in the future for both host and new protist species.

CHAPTER 3

Suborder Pleurodira, Side-Necked Turtles

The Pleurodira are called side-necked turtles because of the method they use to withdraw their heads into their shells. Instead of lowering their necks to pull their heads straight back into the shells, they fold their necks sideways along the body under their shells' margins; this bends their neck in the horizontal plane and draws the head inside their shell tucking it into the space in front of one of the front legs. With the head withdrawn, some of the neck always remains exposed. This differs from the method employed by cryptodiran turtles noted in Chapter 2, which bend the neck in the vertical plane, hence pulling the head straight back between the front legs and completely hiding the neck.

The different methods of bending the neck require a completely different anatomy of the cervical vertebrae. All turtles have seven vertebrae in the neck, as do most higher vertebrates, including mammals. In the Pleurodira, however, these bones are narrow in cross section, they are spool shaped and largely similar to those of other reptiles. They allow for a large degree of sideways movement, but little up and down movement. This division represents a very deep evolutionary divide between two very different types of turtles. The physical differences between them are both anatomical and significant, and the zoogeographic implications are substantial; that is, the Pleurodira turtles are restricted to the Southern Hemisphere, largely to Australia, South America, and Africa.

FAMILY CHELIDAE, AUSTRO-AMERICAN SIDENECK TURTLES, 14 GENERA, 52 SPECIES

Genus *Acanthochelys* Gray, 1873 (4 Species)

To our knowledge, there are no coccidia described from this genus.

Genus *Chelodina* Fitzinger, 1826 (12 Species)

To our knowledge, there are no coccidia described from this genus.

The Biology and Identification of the Coccidia (Apicomplexa) of Turtles of the World.
DOI: http://dx.doi.org/10.1016/B978-0-12-801367-0.00003-4

Genus *Chelus* Schneider, 1783 (Monospecific)

To our knowledge, there are no coccidia described from this genus.

Genus *Elseya* Gray, 1867 (5 Species)

To our knowledge, there are no coccidia described from this genus.

Genus *Elusor* Cann & Legler, 1994 (Monospecific)

To our knowledge, there are no coccidia described from this genus.

Genus *Emydura* Bonaparte, 1836 (4 Species)

To our knowledge, there are no coccidia described from this genus.

Genus *Hydromedusa* Wagler, 1830 (2 Species)

To our knowledge, there are no coccidia described from this genus.

Genus *Mesoclemmys* Gray, 1873 (10 Species)

Eimeria jirkamoraveci Široký et al. (2006a)

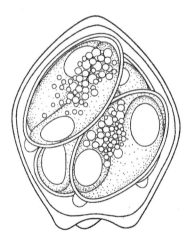

Figures 3.1, 3.2 Line drawing of the sporulated oocyst of Eimeria jirkamoraveci *from Široký et al. (2006a), with permission from the* Memorias do Instituto Oswaldo Cruz *and from Široký and Modrý. Photomicrograph of a sporulated oocyst of* E. jirkamoraveci *from Široký et al. (2006a), with permission from the* Memorias do Instituto Oswaldo Cruz *and from Široký and Modrý.*

Type host: *Mesoclemmys heliostemma* (McCord, Joseph-Ouni & Lamar, 2001) (syn. *Batrachemys heliostemma* McCord, Joseph-Ouni & Lamar, 2001), Amazon Toadhead Turtle.

Type locality: SOUTH AMERICA: Peru, Departmento Loreto, Iquitos, Anquilla Village (03° 54′ 45″ S, 073° 39′ 39″ W).

Other hosts: None to date.

Geographic distribution: SOUTH AMERICA: Peru.

Description of sporulated oocyst: Oocyst shape: ovoidal to subspheroidal to almost spheroidal; number of walls: 1; wall characteristics: smooth, colorless, ~0.5 thick, but one end of the oocyst is conically stretched, while the opposite end bears three blunt conical tubercles, giving the oocyst a mitra-like appearance; L × W ($N = 30$): 10.6 × 8.9 (8−12 × 7−10); L/W ratio: 1.2 (1.0−1.5); M, OR, PG: all absent. Distinctive features of oocyst: sporocysts are packed tightly within the confines of the oocyst wall.

Description of sporocyst and sporozoites: Sporocyst shape: elongate ellipsoidal; L × W ($N = 30$): 7.2 × 4.1 (6−8 × 4−4.5); L/W ratio: 1.75 (1.5−2.0); SB: present, small, knob like, ~0.5−1.0 high- × 1.0−1.5 wide; SSB: probably absent as it was not discernible; PSB: absent; SR: present; SR characteristics: composed of fine granules usually organized into a compact spheroid, ~2−4 × 2−3 ($N = 10$) or scattered among the SZ; SZ: sausage shaped, arranged head-to-tail and each SZ has a spheroidal to subspheroidal RB, ~1.5−2 × 2−2.5 ($N = 8$); N: not visible. Distinctive features of sporocyst: small SB, SR granules, and SZ with only one RB.

Prevalence: Unknown; the authors (Široký et al., 2006a) found these oocysts when they examined a mixed fecal sample that originated from two host animals.

Sporulation: Unknown.

Prepatent and patent periods: Unknown.

Site of infection: Unknown.

Endogenous stages: Unknown.

Cross-transmission: None to date.

Pathology: Unknown.

Materials deposited: Photosyntypes are deposited in the Department of Parasitology, University of Veterinary and Pharmaceutical Sciences Brno, Czech Republic, number R 167/02.

Remarks: This species is dissimilar from all *Eimeria* species parasitizing side-necked turtles. Only two *Eimeria* with mitra-like oocysts resemble this form described by Široký et al. (2006a), *Eimeria mitraria* and *Eimeria stylosa*. Oocysts of *E. stylosa* are much larger and have much longer and sharper projections than those of *E. jirkamoraveci*, while those of *E. mitraria* are the most similar to this species; thus, the authors (Široký et al., 2006a) gave it particular attention. In the original description by Laveran and Mesnil (1902), they reported oocysts of *E. mitraria* to be 15×10, but those of *E. jirkamoraveci* are smaller and more spheroidal with less pronounced conical projections of the oocyst wall. Based on these slight morphological differences, geographic isolation, and the phylogenetic distance of *M. heliostemma* to any other turtle species known to be parasitized by *Eimeria* with mitra-shaped oocysts they (2006a) decided this form was separate and distinct.

Genus *Myuchelys* Thomson & Georges, 2009 (5 Species)
To our knowledge, there are no coccidia described from this genus.

Genus *Phrynops* Wagler, 1830 (4 Species)
To our knowledge, there are no coccidia described from this genus.

Genus *Platemys* Schneider, 1792 (Monospecific)
To our knowledge, there are no coccidia described from this genus.

Genus *Pseudemydura* Siebenrock, 1901 (Monospecific)
To our knowledge, there are no coccidia described from this genus.

Genus *Rheodytes* Legler & Cann, 1980 (Monospecific)
To our knowledge, there are no coccidia described from this genus.

Genus *Rhinemys* Wagler, 1830 (Monospecific)
To our knowledge, there are no coccidia described from this genus.

SUPERFAMILY PELOMEDUSOIDEA

FAMILY PELOMEDUSIDAE, AFRO-AMERICAN SIDENECK TURTLES, 2 GENERA, 19 SPECIES

Genus *Pelomedusa* Lacépède, 1788 (Monospecific)
Eimeria lokuma Široký et al. (2006b)
Type host: *Pelomedusa subrufa* (Bonnaterre, 1789) (syn. *Testudo subrufa* Lacépède, 1788), Helmeted Turtle or Marsh Terrapin.
Type locality: AFRICA: Kenya, between Tangulbei and Lake Baringo (0° 43′ 37″ N, 36° 01′ 03″ E).
Other hosts: None to date.
Geographic distribution: AFRICA: Kenya.
Description of sporulated oocyst: Oocyst shape: subspheroidal to spheroidal; number of walls: 2; wall characteristics: smooth, colorless, ~ 0.7

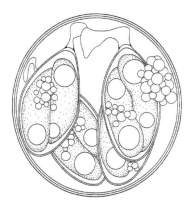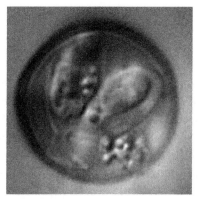

Figures 3.3, 3.4 Line drawing of the sporulated oocyst of Eimeria lokuma *from Široký et al. (2006b), with permission from* Systematic Parasitology *and from Široký and Modrý. Photomicrograph of a sporulated oocyst of* E. lokuma *from Široký et al. (2006b), with permission from* Systematic Parasitology *and from Široký and Modrý.*

(0.5−0.8) thick; L × W (*N* = 22): 13.6 × 13.0 (13−14.5 × 12−14); L/W ratio: 1.05 (1.0−1.2); M, PG: both absent; OR: present; OR characteristics: globular, 3 × 3.5 (2.5−4 × 3−4.5), composed of fine, granular material. Distinctive features of oocyst: mostly spheroidal shape and an OR of fine, granular matter.

Description of sporocyst and sporozoites: Sporocyst shape: elongate ovoidal to spindle shaped; L × W (*N* = 24): 8.3 × 4.4 (7.5−9.5 × 4−5); L/W ratio: 1.9 (1.6−2.1); SB: present as a low, flat structure, ∼0.3−0.5 high × 1.2−1.5 wide, with a flexible, membranous, scarf-like body that covers it; SSB: likely absent and was not discernible; PSB: absent; SR: present; SR characteristics: granular material scattered among the SZ; SZ: sausage shaped, slightly pointed at one end, arranged head-to-tail, and each SZ has spheroidal to subspheroidal RBs, ∼1.5−2 × 2−2.5, at both ends; N: was not discernible. Distinctive features of sporocyst: a small, flat SB with a flexible, membranous, scarf-like body that covers it.

Prevalence: In 1/9 (11%) host animals.

Sporulation: Široký et al. (2006b) stated that sporulation was "probably exogenous." However, they had no way to know this since feces with oocysts were placed into 2.5% aqueous (w/v) potassium dichromate solution ($K_2Cr_2O_7$) when collected in Kenya and the samples were not examined microscopically until transported to their lab in the Czech Republic, at which time, "examined samples contained only fully-sporulated oocysts."

Prepatent and patent periods: Unknown.

Site of infection: Unknown.

Endogenous stages: Unknown.

Cross-transmission: None to date.

Pathology: Unknown.

Materials deposited: Photosyntypes are deposited in the Department of Parasitology, University of Veterinary and Pharmaceutical Sciences Brno, Czech Republic, number R 91/05.

Entymology: The specific epithet "lokuma" is adopted from the Masai language, which uses this vernacular name both for freshwater turtles and land tortoises (Spawls et al., 2002). It was given by Široký et al. (2006b) in accordance with the International Code of Zoological Nomenclature (Article 31.1) as a noun in apposition (ICZN, 1999).

Remarks: Nearly 70 species of eimeriid coccidia have been named and described from turtles worldwide and this diversity has led to overlap in the limited number of morphological characters used as their principle diagnostic features (Duszynski & Wilber, 1997; Široký et al., 2006b). Thus, (presumed) host specificity, host systematic relationships, and geographic origin of the hosts also are used commonly in the taxonomy of the coccidia. No coccidian species had previously been described from this family and genus or from this part of Africa prior to the work of Široký et al. (2006b), who thus defined *E. lokuma* as a new species.

Genus *Pelusios* Wagler, 1830 (18 Species)

To our knowledge, there are no coccidia described from this genus.

FAMILY PODOCNEMIDIDAE, MADAGASCAN BIG HEADED, AND AMERICAN SIDENECK RIVER TURTLES, 3 GENERA, 8 SPECIES

Genus *Erymnochelys* Grandidier, 1867 (Monospecific)

To our knowledge, there are no coccidia described from this genus.

Genus *Peltocephalus* Duméril & Bibron, 1835 (Monospecific)
Eimeria peltocephali Lainson & Naiff (1998)

Type host: *Peltocephalus dumerilianus* (Schweigger, 1812), Big-headed Amazon River Turtle.

Type locality: SOUTH AMERICA: (north) Brazil, State of Amazonas, Barcelos (0.58′ S: 62.57′ W)

Other hosts: Unknown.

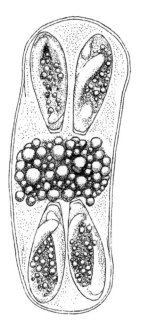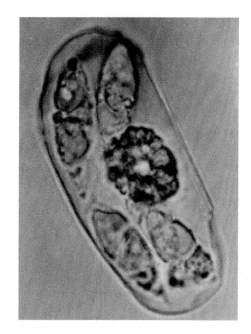

Figures 3.5, 3.6 Line drawing of the sporulated oocyst of Eimeria peltocephali *from Lainson and Naiff (1998), with permission from the* Memorias do Instituto Oswaldo Cruz *and from the senior author. Photomicrograph of a sporulated oocyst of* E. peltocephali *from Lainson and Naiff (1998), with permission from the* Memorias do Instituto Oswaldo Cruz *and from the senior author.*

Geographic distribution: SOUTH AMERICA: (north) Brazil, State of Amazonas.

Description of sporulated oocyst: Oocyst shape: elongate, frequently in the form of a gently curved cylinder with rounded ends; number of walls: 1; wall characteristics: colorless layer, ~1 thick; L × W: 54.4 × 19.1 (37.5−69 × 19−20); L/W ratio: 2.8 (1.8−3.9); M: absent; OR: present; OR characteristics: at first a semi-spheroidal mass of globules, 19 × 16 (16−26 × 16−21.5) in the unsporulated oocyst, but usually dispersed upon sporulation; PG: absent. Distinctive features of oocyst: cylindroidal shape, extreme length, and large L/W ratio.

Description of sporocyst and sporozoites: Sporocyst shape: walls very delicate, frequently located in pairs toward the ends of the oocyst; L × W: 19.1 × 6.8 (17.5−21 × 6−7.5); L/W ratio: 2.8 (2.3−3.2); SB: present as an inconspicuous cap-like structure; SSB, PSB: both absent; SR: present; SR characteristics: bulky, composed of a mixture of fine and large granules; SZ: occupy almost the entire length of the sporocyst and are slightly recurved around the

SR; neither RBs nor N were visible to the original authors. Distinctive features of the sporocysts: they usually occur in pairs at either end of the oocyst separated by the SR.

Prevalence: Found in 9/18 (50%) specimens of the type host.

Sporulation: Presumably exogenous, but the sporulation time was not recorded (Lainson & Naiff, 1998).

Prepatent and patent periods: Unknown.

Site of infection: Uncertain, but with the failure to detect parasites in the gallbladder of infected animals, it was "most probably in the intestine" (Lainson & Naiff, 1998). Oocysts were found only in the feces.

Endogenous stages: Unknown.

Cross-transmission: None to date.

Pathology: Unknown.

Materials deposited: None.

Entymology: The specific epithet was derived from the generic name of the host, *Peltocephalus* (Lainson & Naiff, 1998).

Remarks: Lainson and Naiff (1998) provided a reasonably complete table of 45 of the *Eimeria* species recorded from turtles through about 1997. Although they had missed a few pertinent citations (e.g., *Eimeria lecontei* Upton et al., 1995; *Eimeria geochelona* Couch et al., 1996), their review of most of the literature gave them and others strong assurance that *E. peltocephali* was, in fact, distinct from other eimerians described from turtles to that date. We encourage all others who do alpha taxonomy, or other taxonomic reconstructions, to serve their future readers and colleagues by being so thorough. The only two other eimerians with elongate–cylindroidal oocysts noted in the literature are *Eimeria texana* and *Eimeria cooteri* (McAllister & Upton, 1989b), but both sporulated oocysts are much smaller (*E. peltocephali* is 54.4 × 19.1, L/W 2.8 vs. *E. texana*, 20.5 × 8.4, L/W 2.4 vs. *E. cooteri*, 25.9 × 10.9, L/W 2.4), and there are numerous differences between the sporocysts of the three species.

Genus *Podocnemis* Wagler, 1830 (6 Species)
Eimeria lagunculata Lainson et al. (1990)

Type host: *Podocnemis expansa* (Schweigger, 1812), Arrau or South American river turtle.

Type locality: SOUTH AMERICA: (north) Brazil: Pará, Belém; turtles housed in the Museu Paraense Emilio Goeldi.

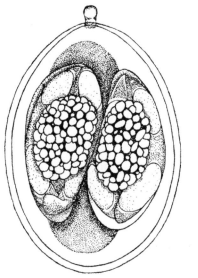 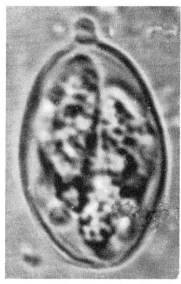

Figures 3.7, 3.8 Line drawing of the sporulated oocyst of Eimeria lagunculata *from Lainson et al. (1990), with permission from the* Memorias do Instituto Oswaldo Cruz *and from the senior author. Photomicrograph of a sporulated oocyst of* E. lagunculata *from Lainson et al. (1990), with permission from the* Memorias do Instituto Oswaldo Cruz *and from the senior author.*

Other hosts: None to date.

Geographic distribution: SOUTH AMERICA: (north) Brazil: Pará, Belém.

Description of sporulated oocyst: Oocyst shape: ellipsoidal; number of walls: 1; wall characteristics: very delicate, smooth, colorless, and a single layer, ~0.5−0.7 thick; L × W (N = 50): 19.2 × 12.8 (17−21 × 12−14); L/W ratio: 1.5 (1.4−1.7); M: present, conspicuous, at one pole of the oocyst as a stopper-like structure, ~1.5−2.0 long × 1.0−1.5 wide; OR, PG: both absent. Distinctive features of oocyst: the distinctive stopper-like M.

Description of sporocyst and sporozoites: Sporocyst shape: elongate−ellipsoidal, often with one side flattened and a very delicate wall; L × W (N = 50): 11.0 × 5.4 (10−12 × 5−6); L/W ratio: 2.0 (1.8−2.1); SB, SSB, PSB: all absent; SR: present; SR characteristics: compact, usually ellipsoidal, and composed of numerous small granules lying between the two sporozoites. SZ: present with posterior and anterior RBs, which are seen with difficulty. Distinctive features of sporocyst: flattened on one side and lacking SB, SSB, and PSB.

Prevalence: Unknown in the wild; 5/7 (71%) turtles examined in the Museu Paraense Emilio Goeldi, Belém, were infected. Of these, one was infected only with *E. lagunculata*, one with both *E. lagunculata* and *E. podocnemis*, and three were infected with *E. lagunculata* and *E. mammiformis*.

Sporulation: Endogenous; both sporulated and developing oocysts were found in fecal material examined immediately after removal from the intestine.

Prepatent and patent periods: Unknown.

Site of infection: The ileum; no parasites were seen in the gallbladder contents, liver, or spleen. Oocysts collected from the feces were described.

Endogenous stages: Unknown.

Cross-transmission: None to date.

Pathology: Infected turtles appeared to be healthy.

Materials deposited: Sporulated oocysts were preserved in 10% formal saline and are held in the Department of Parasitology, Instituto Evandro Chagas, Belém, Pará, Brazil.

Entymology: The specific epithet is from the Latin *laguncula* (a little flask), suggesting the flask-like appearance given to the oocyst by the M.

Remarks: The strange "stopper-like" M of *E. lagunculata* differentiates the oocysts of this form from all the eimeriid species described from chelonids to date. Of the 28 that Lainson et al. (1990) found in the available literature to that date, only five had some form of M; in no case, however, was the M of any other species comparable with that of *E. lagunculata*, because they were manifest as merely a thinning of their oocyst walls at one point or, at most, a flat cap-like body. Apart from this difference in the micropyle structure, the other species with an M are readily differentiated from *E. lagunculata* as follows. Oocysts of *Eimeria brodeni*, of the Greek tortoise *Testudo graeca*, has larger oocysts which are oval in shape, measure 28–32 × 18–20, and have an oocyst wall with two layers. Those of *Eimeria chrysemydis*, described from the North American turtle, *Chrysemys picta marginata*, has larger, pear-shaped oocysts, 23.0 × 15.0 (Deeds & Jahn, 1939) or 27.6 × 17.0 (Wacha & Christiansen, 1976), and a yellow, two-layered oocyst wall. Oocysts of *Eimeria koormae*, of the Indian tortoise *Lissemys punctata*, are readily distinguished by being spheroidal, averaging

14.0 wide, and having more elongated sporocysts (L/W = 2.2). Those of *Eimeria marginata*, from *Chrysemys picta marginata*, have larger, pear-shaped oocysts, 20–28 × 15–21 and larger, elongated sporocysts, 10–14 × 15–21, L/W = 1.5. Finally, oocysts of *Eimeria scriptae* from *Pseudemys scripta elegans*, also from North America, are distinguished by their larger size, 24.2 × 13.7, and having two to three layers in their oocyst wall.

Eimeria mammiformis Lainson et al. (1990)

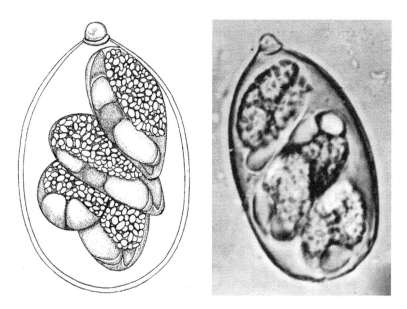

Figures 3.9, 3.10. Line drawing of the sporulated oocyst of Eimeria mammiformis *from Lainson et al. (1990), with permission from the* Memorias do Instituto Oswaldo Cruz *and from the senior author. Photomicrograph of a sporulated oocyst of* E. mammiformis *from Lainson et al. (1990), with permission from the* Memorias do Instituto Oswaldo Cruz *and from the senior author.*

Type host: *Podocnemis expansa* (Schweigger, 1812), Arrau or South American river turtle.
Type locality: SOUTH AMERICA: (north) Brazil: Pará, Belém; turtles housed in the Museu Paraense Emilio Goeldi.
Other hosts: None to date.
Geographic distribution: SOUTH AMERICA: (north) Brazil: Pará, Belém.
Description of sporulated oocyst: Oocyst shape: ellipsoidal, narrowing at one pole with the M; number of walls: 1; wall characteristics: smooth, colorless, ~0.7 thick; L × W (*N* = 50): 30.0 × 19.4

(23−37 × 16−21.5); L/W ratio: 1.5 (1.1−1.9); M: present as a nipple-like structure at the narrow end of the oocyst, ∼2.5−3.0 high × 3.0 wide; OR: absent; PG: rarely present as a single body, 1−2 × 0.5−1.5. Distinctive features of oocyst: Unique nipple-like M and relatively large size.

Description of sporocyst and sporozoites: Sporocyst shape: elongate−ellipsoidal, slightly pointed at one end, frequently with one side flattened and a very delicate wall; L × W (*N* = 50): 15.3 × 7.9 (15−17 × 7−10); L/W ratio: 2.0 (1.8−2.2); SB: present as a minute body on the wall of the sporocyst; SSB, PSB: both absent; SR: present; SR characteristics: bulky, usually as an ellipsoidal mass of relatively large granules, which often obscure the SZ; SZ: liberated SZ were 14.8 × 3.7 in fresh preparations and each has a large posterior RB and a smaller anterior one. Distinctive features of sporocyst: flattened on one side, presence of a minute SB, and a bulky SR with a mass of large granules that often obscure the SZ.

Prevalence: Unknown in the wild; 4/7 (57%) captive turtles examined were infected, but all were housed in the same aquarium. In three cases the animals had mixed infections with *E. mammiformis* and *E. lagunculata*.

Sporulation: Endogenous; mature and developing oocysts were present in fecal material examined immediately after its removal from the intestine.

Prepatent and patent periods: Unknown.

Site of infection: The small intestine (ileum) was verified by intestinal scrapings. No parasites were seen in gallbladder contents or in the liver or spleen.

Endogenous stages: Unknown.

Cross-transmission: None to date.

Pathology: Infected turtles appeared healthy.

Materials deposited: Sporulated oocysts were preserved in 10% formal saline and are held in the Department of Parasitology, Instituto Evandro Chagas, Belém, Pará, Brazil.

Entymology: The specific epithet is from the Latin *mamma* (breast) and *forma* (shape) to reflect the nipple-like M.

Remarks: No *Eimeria* species possessing a micropyle comparable with that of *E. mammiformis* has been described previously from chelonid hosts (also see Remarks for *E. lagunculata*, above).

Eimeria podocnemis Lainson et al. (1990)

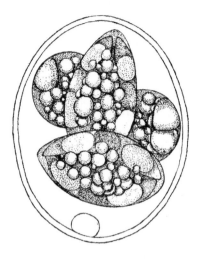
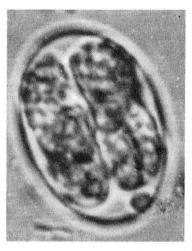

Figures 3.11, 3.12 Line drawing of the sporulated oocyst of Eimeria podocnemis from Lainson et al. (1990), with permission from the Memorias do Instituto Oswaldo Cruz and from the senior author. Photomicrograph of a sporulated oocyst of E. podocnemis from Lainson et al. (1990), with permission from the Memorias do Instituto Oswaldo Cruz and from the senior author.

Type host: *Podocnemis expansa* (Schweigger, 1812), Arrau or South American river turtle.

Type locality: SOUTH AMERICA: (north) Brazil: Pará, Belém; turtles housed in the Museu Paraense Emilio Goeldi.

Other hosts: None to date.

Geographic distribution: SOUTH AMERICA: (north) Brazil: Pará, Belém.

Description of sporulated oocyst: Oocyst shape: broadly ellipsoidal, with rounded extremities; number of walls: 1; wall characteristics: smooth, colorless, ~0.5–0.7 thick; L × W ($N = 50$): 17.0 × 12.8 (15–19 × 12–14); L/W ratio: 1.3 (1.1–1.4); M, OR: both absent; PG: present as a single, pale body, ~2.0 × 1.5, always seen attached to the inside layer of the wall at one end of all oocysts examined. Distinctive features of oocyst: PG attached to the inside layer of the wall at one pole of all oocysts.

Description of sporocyst and sporozoites: Sporocyst shape: ellipsoidal, with both ends tending to be somewhat pointed and very delicate; L × W ($N = 50$): 9.7 × 5.2 (9–10 × 4–6); L/W ratio: 1.9 (1.6–2.0); SB, SSB, PSB: all absent; SR: present; SR characteristics: ellipsoidal mass of 20–25 large and medium round globules that obscure the SZs; SZ: each has a posterior and anterior RB, but they are seen with difficulty. Distinctive features of sporocyst: massive SR that mostly fills the sporocyst and obscures the SZ.

Prevalence: Unknown in the wild. Only 1/7 (14%) captive turtles examined was infected, and this animal also was infected with *E. lagunculata*.

Sporulation: Endogenous; mature and developing oocysts seen in fecal material immediately after its removal from the intestine.

Prepatent and patent periods: Unknown.

Site of infection: Small intestine (ileum); no parasites were seen in gallbladder contents or in the liver and spleen.

Endogenous stages: Unknown.

Cross-transmission: None to date.

Pathology: Presumably none, as the infected turtle showed no signs of ill-health.

Materials deposited: Sporulated oocysts were preserved in 10% formal saline and are held in the Department of Parasitology, Instituto Evandro Chagas, Belém, Pará, Brazil.

Entymology: The specific epithet is derived from the generic name of the type host, *Po. expansa*, in which this parasite was first found.

Remarks: There is no recorded species of *Eimeria* which combine the oocyst size range and other morphologic features of *E. podocnemis*. The species with oocysts that resemble it most closely, *Eimeria jaboti*, shares or lacks many of the same features; they both lack an M and an OR and have a single PG and sporocysts with no SB. However, oocysts of *E. jaboti* are mostly spheroidal, 17.0 wide, to subspheroidal, $17-19 \times 15-17$, and have a relatively thick, two-layered wall and their ovoidal sporocysts are only $10-11 \times 6-7$, and sporulation is exogenous. These structural differences of their oocysts along with host genus and familial differences were reasons to distinguish this form from *E. jaboti*.

DISCUSSION AND SUMMARY

Members of suborder Pleurodira are distinct from members of suborder Cryptodira in having necks that fold sideways into their shells as opposed to having necks that retract. The lineage of this suborder can be traced back to the Cretaceous (Georges et al., 1998). This group of turtles is much less speciose than is the Cryptodira (Chapter 2); it consists of only three families with 19 genera that contain 79 species.

At least one species in each of the three families has been examined for coccidia, but only four of the 19 (21%) turtle genera and four of the 79 (5%) species in this suborder have been examined for coccidia. From these four host species, six *Eimeria* species have been discovered, described, and named.

In the Chelidae (Austro-American Sideneck turtles), only *E. jirkamoraveci* is known from *M. heliostemma*, from Peru, South America. In the Pelomedusidae (Afro-American Sideneck turtles), only *E. lokuma* is known from *Pe. subrufa*, from Kenya, Africa. And in the Podocnemididae (Madagascan big-headed turtles and American sideneck river turtles), four eimerians, all from northern Brazil, South America, are known from two turtle genera: *E. peltocephali* has been described from *Pe. dumerilianus*; and *E. lagunculata*, *E. mammiformis*, and *E. podocnemis* from *Po. expansa*.

Within the Cryptodira (Chapter 2), mainly because so much more work has been done on species in that suborder, we were able to comment on coccidiosis, prevalence, endogenous development, host specificity, and the topic of archiving biological host and parasite specimens. So what do we know about these topics and the general biology of the six *Eimeria* species in the Pleurodira, other than the morphometrics of their sporulated oocysts?

Specimens have been deposited in various museums for five of the six eimerians, either as photosyntypes of sporulated oocysts or sporulated oocysts preserved in formalin (90–100% ethanol would have been a much wiser choice). Three of the six eimerians are thought to begin sporulation endogenously because both sporulated and unsporulated oocysts were found in fecal material either taken directly from the intestine or examined immediately after removal from the intestine. Three of the six eimerians were speculated to undergo endogenous development in the ileum. However, it was not clear from any of the original descriptions whether developmental stages were *actually seen* within cells of the ileum in fresh tissue squash preparations or whether the statements made by authors who described them only refers to the *presumed* location of merogony and gamogony because no parasites were seen in gallbladder contents or in the liver. Fixed tissue sections were not examined for any of these turtles from which eimerians have been described to date. This short paragraph comprises the extent of our knowledge on these parasites.

Other than the type hosts from which the six *Eimeria* species were described, and their type localities on two southern hemisphere continents (Kenya, Africa; Brazil and Peru, South America), we know nothing else. We don't know if other sympatric turtle species can or do harbor them. We don't know anything about their host ranges. We know little or nothing about their prevalence in wild populations, both because so few host animals have been sampled and some were sampled from several captive animals maintained in the same aquarium. We know nothing about the prepatent and patent periods and no descriptions exist for any stages of merogony or gamogony. There has been no cross-transmission work done on any turtle hosts to date, so we have no information on host specificity, and because surveys have been so few and sample sizes so small, we don't know if multiple host species can become infected with any of these species. None of these six turtle eimerians are thought to cause pathology, but that is only because "infected turtles appeared healthy" whereas, in reality, we do not know whether turtle eimerians can produce pathology or not in these hosts.

CHAPTER 4

Cryptosporidium, Sarcocystis, Toxoplasma in Turtles

CRYPTOSPORIDIUM IN TURTLES

Cryptosporidium has been identified from at least the following turtle species: *Chelonia mydas* (L., 1758), Green Turtle (Graczyk et al., 1997); *Chelonoidis carbonaria* (Spix, 1824) (syn. *Geochelone carbonaria* Ernst & Barbour, 1989), Red-footed Tortoise (Funk, 1987); *Cuora amboinensis* (Daudin, 1802), Amboina Box Turtle (Pavlásek, 1998); *Geochelone elegans* (Schoepff, 1795), Star Tortoise (Heuschele et al., 1986; Graczyk & Cranfield, 1998); *Glyptemys muhlenbergi* (Schoepff, 1801) (syn. *Clemmys muhlenbergii* Fitzinger, 1835), Bog Turtle (Graczyk et al., 1996; Graczyk & Cranfield, 1998); a *Kinixys* sp. from Africa imported into France (Bourdeau, 1988, 1989); and *Testudo kleinmanni* Lortet, 1883, Egyptian Tortoise (Graczyk et al., 1988). We do not know whether these forms represent one of more species and, for the moment, they must all be considered *species inquirendae*.

Cryptosporidium parvum, one of about 18 valid, named species (Duszynski & Upton, 2009), infects humans and is a global public health threat, having caused massive waterborne epidemics (Graczyk et al., 1997). Graczyk et al. (1997) were the first to detect *Cryptosporidium* oocysts in fecal and intestinal samples from free-ranging marine green turtles, *Ch. mydas*. Their intent was to determine whether or not these oocysts would produce positive reactions with commercial test kits that had been recommended for the detection of human-infectious waterborne oocysts of *Cr. parvum*. Six of the 34 (18%) turtles they examined had *Cryptosporidium* oocysts. All isolates produced positive direct- and indirect-IFA (Immunofluorescent Assay) reactions, but none produced a positive EIA (Enzyme immunoassay) reaction indicating that these oocysts did not represent *Cr. parvum*. Their turtle-recovered oocysts were bigger than those of *Cr. parvum* and their excystation rate and excystation index were lower than those of *Cr. parvum*. They cautioned that if the test kits they used, or similar ones, were used by public health agencies and reacted positively with

The Biology and Identification of the Coccidia (Apicomplexa) of Turtles of the World.
DOI: http://dx.doi.org/10.1016/B978-0-12-801367-0.00004-6

oocysts from turtles (or other animals) the result can be misleading and would needlessly exacerbate public health concerns. The molecular similarities of turtle-recovered oocyst to *Cr. parvum*, and their infectivity for immune-compromised or immunosuppressed people are not known (Graczyk et al., 1997).

Graczyk et al. (1998) gave us the first histological documentation of intestinal cryptosporidiosis in turtles, when they found an Egyptian tortoise (*T. kleinmanni*) with obvious clinical signs of enteritis and necropsied it when it died 5 days later. They found no developmental stages in the stomach or lungs, but the intestinal lamina propria and mucosa were heavily infected with developmental stages and oocysts, with >80% of epithelial cells harboring these stages. They also noted massive infiltration of the lamina propria and mucosa by heterophils, lymphocytes, and macrophages.

Pavlásek (1998) found *Cryptosporidium* oocysts in the watery feces of an Amboina box turtle, *Cu. amboinensis*, which came from Vietnam. He found numerous, small-form oocysts that measured 4.8 × 4.8 (4.8−5.1 × 4.8), and said that these oocysts very much resembled those of *Cryptosporidium varanii*, the agent of intestinal cryptosporidiosis in *Varanus prasinus*, the Emerald monitor.

Häfeli and Zwart (2000) observed weakening of the carapace in young tortoises (*Testudo* spp.) and, based on histopathology, the problem was diagnosed as a hypoplastic osteoporosis in juveniles. They suggested that possible causes were intestinal infections with endoparasites such as *Cryptosporidium* and *Balantidium* species.

Most of our information about host specificity (or lack thereof) on *Cryptosporidium* species is based on the data from mammals. Thus, it is believed that members of this genus are *not* very host specific and can infect numerous hosts. For example, in other reptiles that have been studied in some detail (see Duszynski & Upton, 2009) there are two species that are accepted to parasitize reptiles, one in snakes (*Cryptosporidium serpentis*) and one in lizards (*Cryptosporidium saurophilus*), but many genera and species in each group seem to be infected with these forms (Greiner, 2003). Although there have been a good number of *Cryptosporidium* reports from turtles (e.g., Graczyk et al., 1997, 1998; Pavlásek, 1998; Häfeli and Zwart, 2000; others) no one yet has attempted to delineate the complete life cycle, to PCR any gene fragments, or to name one. Realistically, we know almost nothing about *Cryptosporidium* in turtles (and other reptiles) except that they are there. This certainly would be a fruitful and rewarding area for future research.

SARCOCYSTIS IN TURTLES

We know even less about *Sarcocystis* infections in turtles than we do about *Cryptosporidium*. Only seven turtle species representing three genera have been documented to have tissue cysts of *Sarcocystis*: *Ch. carbonaria* (Spix, 1824) (syn. *Testudo carbonaria* Spix, 1824), Red-footed Tortoise (Weishaar et al., 1988); *Chelonoidis denticulata* (L., 1766) (syn. *Testudo denticulata* L., 1766), Yellow-footed Tortoise (Keymer, 1978a,b); *Kinosternon scorpiodes* (L., 1766), Scorpion Mud Turtle (Lainson & Shaw, 1971, 1972); *Testudo graeca* L., 1758, Spur-thighed Tortoise (Meshkov, 1975; Weishaar et al., 1988); *Testudo hermanni* Gmelin, 1789, Hermann's Tortoise (Meshkov, 1975; Weishaar et al., 1988); *Testudo horsfieldii* Gray, 1844, Horsfield's Tortoise; and *Testudo marginata* Schoepff, 1792, Marginated Tortoise. With the exception of the reports by Lainson and Shaw (1971, 1972) from *K. scorpiodes*, and the report from Bulgaria (Meshkov, 1975), all of the other turtle species had compartmented cysts with smooth walls and could not be distinguished morphologically.

Meshkov (1975) surveyed muscle tissue from the neck, hind legs, and front legs of 49 tortoises (*T. graeca*, *T. hermanii*) from two mountainous areas of Bulgaria. He found "microcysts" in 43/49 (88%) turtles, but provided no measurements, descriptions, drawings, or photographs.

There has only been enough work on the tissue cysts from one of these hosts to merit a species name, which we summarize here.

Sarcocystis kinosterni Lainson & Shaw, 1972
Synonym: *Sarcocystis gracilis* Lainson & Shaw, 1971, preoccupied.
Definitive type host: Unknown.
Type locality: SOUTH AMERICA: Brazil: Pará State, Island of Marajó.
Other definitive hosts: Unknown.
Intermediate type host: *Kinosternon scorpioides* (L., 1766), Scorpion Mud Turtle.
Other intermediate hosts: Unknown.

Geographic distribution: SOUTH AMERICA: Brazil.
Description of sporulated oocyst: Unknown.
Description of sporocyst and sporozoites: Unknown.
Prevalence: Muscle cysts were found in 18/206 (9%) *K. scorpioides*.

Sporulation: Unknown, but likely endogenous with infective sporocysts shed in the feces of the definitive host.
Prepatent and patent periods: Unknown.
Site of infection, definitive host: Unknown, presumably the epithelium of the small intestine.
Site of infection, intermediate host: Skeletal muscle fibers, but no cysts were found in heart muscle, involuntary muscle of the intestine, viscera, or brain.
Endogenous stages, definitive host: Unknown.
Endogenous stages, intermediate host: Sarcocysts were cylindrical with pointed ends, up to 8 mm long and ∼230 μm wide in fresh preparations; relatively smooth outline, only rarely with invaginations or convolutions. Trabeculae were well marked and divided the cyst into angular compartments containing up to many hundreds of zoites. Zoites were long, slender, and curved into a graceful crescent, up to 18.4 × 1.7 in fresh preparations. Each zoite with an N either in its center or closer to the blunter end. Zoites showed sudden and rapid gliding movements in fresh preparations.
Cross-transmission: None to date.
Pathology: Frequently produced marked pathologic changes in the muscle.
Materials deposited: Slides of stained tissue sections are in the Department of Parasitology, London School of Hygiene and Tropical Medicine, Keppel Street, London, W.C.1 and Wellcome Parasitology Unit, Instituto Evandro Chagas, Belém, Pará, Brazil.
Remarks: This is the first record of a sarcosporidian from turtles and, to our knowledge, the only named species in the literature to date.

TOXOPLASMA IN TURTLES

Although there are numerous reports of *Toxoplasma*-like infections in reptiles, no natural infections have been reported yet from turtles, to our knowledge. Stone and Manwell (1969) used intraperitoneal injection of ascitic fluid, containing *Toxoplasma gondii* zoites from infected mice (RH strain of human origin), and successfully transferred infections to horned toads (*Phrynosoma cornutum*), green anoles (*Anolis carolinensis carolinensis*), blue spiny lizards (*Sceloporus cyanogenys*), red-eared sliders (*Trachemys (Pseudemys) scripta elegans*), and coal skinks (*Eumeces anthrospinus*); infection persisted in all animals for 7–8 days, but "skinks had a fulminating infection which was fatal in a week or less" (Stone & Manwell, 1969). In turtles, 2/13 (15%) sliders maintained infection for up to 7 days duration. The question in these hosts is whether this was a true infection or was it simply survival of the zoites for 7 days? Stone and Manwell (1969) concluded that their results "suggest that natural infections could occur under suitable environmental conditions" in some reptiles.

CHAPTER 5

Species Inquirendae in Turtles

SPECIES INQUIRENDAE (28)

Coccidium sp. of Jacobson et al., 1994

Original host: *Astrochelys radiata* (Shaw, 1802) (syn. *Geochelone radiata* Ernst & Barbour, 1989), Radiated Tortoise.

Remarks: Jacobson et al. (1994) wrote about an intranuclear coccidian and described various tissue stages in two juvenile radiated tortoises maintained in an outdoor enclosure in north-central Florida (that had been derived from a captive breeding group on St. Catherine's Island, GA), but they never found or reported oocysts in the feces and, thus, the generic determination of this coccidium remains unknown. In their paper, they said the tortoises were anorexic and lethargic. Upon histological examination they described enteritis, hepatitis, nephritis, and pancreatitis associated with this coccididal infection (Jacobson et al., 1994). Light microscopic examination showed an intranuclear protozoan in renal epithelial cells, hepatocytes, pancreatic acinar cells, and duodenal epithelial cells. Electron microscopic findings demonstrated to them what they believed was an "intranuclear coccidian parasite." Clearly, the reports by Jacobson et al. (1994), Garner et al. (2006), and Innis et al. (2007) have identified and addressed an important coccidium (-ia?) of turtles that desperately needs additional attention and research funding, especially in that a number of endangered turtle species seem to be infected with this systemic, intranuclear organism.

Coccidium sp. of Garner et al., 1998

Original hosts: *Manouria impressa* (Günther, 1882), Impressed Tortoise and *Stigmochelys pardalis* (Bell, 1828) (syn. *Geochelone pardalis* Bour, 1980), Leopard Tortoise.

Other hosts: *Astrochelys radiata* (Shaw, 1802) (syn. *Geochelone radiata* Ernst & Barbour, 1989), Radiated Tortoise; *Indotestudo forstenii* (Schlegel & Müller, 1844), Travancore or Foresten's Tortoise.

The Biology and Identification of the Coccidia (Apicomplexa) of Turtles of the World.
DOI: http://dx.doi.org/10.1016/B978-0-12-801367-0.00005-8

Remarks: Garner et al. (1998) said that between 1994–1998, 93 chelonians were submitted for examination to N.W. ZooPath, Snohomish, WA, and 2 (2%) were found to be infected with intra-nuclear coccidians; a wild-caught *M. impressa* adult male and a seven-mo-old *S. pardalis*. They also reported that five other tortoises submitted to the Wildlife Conversation Society, Bronx, NY, were infected with a similar nuclear coccidian; these included two adult *A. radiata* from St. Catherine's Island, GA, and three *I. forstenii* taken from an illegal tortoise shipment originating in Celebes (Sulawesi) Island, Indonesia.

Coccidium sp. of Garner et al., 2006

Original host: *Astrochelys radiata* (Shaw, 1802) (syn. *Geochelone radiata* Ernst & Barbour, 1989), Radiated Tortoise.

Other hosts: *Indotestudo forstenii* (Schlegel & Müller, 1844), Travancore or Foresten's Tortoise; *Stigmochelys pardalis* (Bell, 1828) (syn. *Geochelone pardalis* Bour, 1980), Leopard Tortoise; *Chersina angulata* (Schweigger, 1812), South African Bowsprit or Angulate Tortoise; *Manouria impressa* (Günther, 1882), Impressed Tortoise.

Remarks: Garner et al. (2006) described intranuclear coccidiosis his-tologically in two radiated tortoises, three Foresten's tortoises, two leopard tortoises, one bowsprit tortoise, and one impressed tortoise. The tortoises were highly variable in their points of origin ranging from captive-bred animals in private facilities (Baton Rouge, LA; St. Catherine's Island, GA; New York, NY), to Zoo animals (central FL), to wild-caught adults (Thailand, Celebes) that had been confis-cated by the US Fish and Wildlife Service during illegal shipments into the United States. It is also likely that these are the same tor-toises reported in their earlier two studies (Jacobson et al., 1994; Garner et al., 1998). In each animal, the infection was systematic and involved the alimentary, endocrine, integumentary, lymphoid, respiratory, and urogenital systems; lesions were most severe in the pancreas, colon, kidney, and lung. Most endogenous stages including trophozoites, meronts, merozoites, macro- and microgametocytes, and unsporulated oocysts were visualized by both light- and electron microscopy. The coccidiosis associated with the presence of this intranuclear parasite was associated with variable degrees of inflam-mation in all nine tortoises, was considered to be the cause of death for six of them, and was said to be a substantial contributing factor to the cause of death in two others.

Garner et al. (2006) were able to obtain nucleic acid sequences from two of the nine animals (one *I. forstenii*, one *S. pardalis*) and an approximately 350-basepair segment of small subunit (SSU) rRNA was PCR-amplified and the products were compared with known sequences in genBank. Their phylogenetic analysis of the intranuclear coccidial gene sequence, which they described as "novel," led them to conclude that their sequence "does not provide support for the inclusion in either the Sarcocystidae or the Eimeriidae, and this organism may be paraphyletic to these families."

Coccidium sp. of Innis et al., 2007

Original host: Indotestudo forstenii (Schlegel & Müller, 1844), Travancore or Foresten's Tortoise.

Remarks: Innis et al. (2007; and an undated report) identified and characterized both an intranuclear coccidium and *Mycoplasma* spp. from the nasal cavities of five Sulawesian Foresten's tortoises (*I. forstenii*) that were affected by chronic rhinosinusitis and oronasal fistulae. Their study provided the first antemortem diagnosis of intranuclear coccidiosis and the histopathologic and ultrastructural data they presented allowed them to conclude that the intranuclear coccidium in their tortoises was identical to those previously described in tortoises (Jacobson et al., 1994; Garner et al., 2006). Nucleic acid sequence data of a 1624- to 1757-basepair fragment of the 18S SSU rRNA gene identified the coccidian they studied as a novel species. This lent support to the conclusion of Garner et al. (2006) that this intranuclear coccidium of tortoises, at least based on this nucleotide sequence, clusters neither within the Sarcocystidae nor the Eimeriidae. They (2007) conclude their remarks on this intranuclear coccidium by pointing out the obvious, but seldom addressed facts, that captive animals (especially their turtles) "have often been through extensive networks of local collectors, regional wholesalers, importers, and pet stores, and have likely been exposed to many other reptile species and pathogens."

Caryospora sp. of Greiner, 2003

Original host: Astrochelys radiata (Shaw, 1802) (?), Radiated Tortoise.

Remarks: Greiner (2003) provided a general review entitled, "Coccidiosis in reptiles." In that paper he had a photomicrograph of a sporulated oocyst of what he called *Caryospora* sp. that he said came from a radiated tortoise. However, he made no mention of this caryosporan in the body of the paper; that is, he did not name it, or describe its morphology, or make any mention about whether or not it caused any pathology, even though he did address the pathology caused by *Caryospora cheloniae* in green sea turtles, *Chelonia mydas*, in the Cayman Islands.

Cryptosporidium sp. of Bourdeau, 1988, 1989
Original host: *Kinixys* sp., Hingeback Tortoise.
Remarks: Bourdeau (1988, 1989) only mentioned that a *Cryptosporidium* sp. caused digestive disorder in a hingeback tortoise he examined.

Cryptosporidium sp. of Funk, 1987
Original host: *Chelonoidis carbonaria* (Spix, 1824) (syn. *Geochelone carbonaria* Ernst & Barbour, 1989), Red-footed Tortoise.
Remarks: Funk (1987) wrote a paper on cryptosporidiosis in emerald tree boas (*Corallus caninus* L., 1758) and made one statement about a *Cryptosporidium* sp. in this tortoise when he wrote, "I have seen additional species of reptiles, including...*Geochelone carbonaria* (red-footed tortoise) with cryptosporidiosis." Nothing else was mentioned.

Cryptosporidium sp. 1 of Graczyk & Cranfield, 1998
Original host: *Geochelone elegans* (Schoepff, 1795), Star Tortoise.
Remarks: Graczyk and Cranfield (1998) isolated some oocysts of a *Cryptosporidium* sp. from *Ge. elegans*, but did not state the origin of the tortoise. They used these oocysts combined with *Cryptosporidium* sp. oocysts from *Glyptemys muhlenbergi* (see below) to infect four three-mo-old rat snakes (*Elaphe obsoleta* (Yarrow, 1880)), all of which became infected and passed oocysts in their feces. They did not name this species.

Cryptosporidium sp. 2 of Graczyk & Cranfield, 1998
Original host: *Glyptemys muhlenbergi* (Schoepff, 1801) (syn. *Clemmys muhlenbergii* Fitzinger, 1835), Bog Turtle.
Remarks: Graczyk and Cranfield (1998) isolated some oocysts of a *Cryptosporidium* sp. from *Gl. muhlenbergi*, but did not state the origin

of the turtle. They used these oocysts combined with *Cryptosporidium* sp. oocysts from *Ge. elegans* (see above) to infect four three-mo-old rat snakes (*Elaphe obsoleta*), all of which became infected and passed oocysts in their feces. They did not name this species.

Cryptosporidium sp. of Graczyk et al., 1997

Original host: *Chelonia mydas* (L., 1758), Green Sea Turtle.

Remarks: The authors found *Cryptosporidium* oocysts for the first time in the fecal and intestinal contents of free-ranging marine turtles collected from the shores of Oahu (22 turtles) and Maui (12 turtles), the Hawaiian Islands. The oocysts they found produced positive reactions with commercial test kits recommended for the detection of human-infectious waterborne oocysts of *Cryptosporidium parvum*. Six of the 34 (18%) fecal and intestinal samples from immature turtles contained oocysts, five (23% prevalence) from Oahu and one (8% prevalence) from Maui. All oocyst samples produced positive direct- and indirect-IFA reactions, but none of them produced a positive EIA reaction (Pro-Spect test), indicating to the authors that the oocysts did not represent *Cr. parvum*. However, they did not provide photographic or other evidence for the oocysts and they did not name it.

Cryptosporidium sp. of Graczyk et al., 1998

Original host: *Testudo kleinmanni* Lortet, 1883, Egyptian Tortoise.

Remarks: Graczyk et al. (1988) were presented with one adult (300 g) Egyptian tortoise that had clinical signs of enteritis and died 5 weeks after antibiotic therapy. They examined the small intestine histologically and found a heavy infection with a *Cryptosporidium* sp., with >80% of the epithelial cells harboring developmental stages of the parasite. They did not, however, find any developmental stages in the stomach or lungs. Theirs was the first report of *Cryptosporidium* sp. in this host and the first histological documentation of intestinal, rather than gastric, cryptosporidiosis in turtles. They did not name the species.

Cryptosporidium sp. of Häfeli & Zwart, 2000

Original host: *Testudo graeca* L., 1758, Mediterranean Spur-thighed Tortoise.

Remarks: The authors only mention finding an intestinal infection with "Cryptosporidia."

Cryptosporidium sp. of Heuschele et al., 1986

Original host: *Geochelone elegans* (Schoepff, 1795), Star Tortoise.

Remarks: The authors examined fecal samples of wild mammals and reptiles held in captivity at the San Diego Wild Animal Park in California, USA. One star tortoise showed signs of clinical gastritis; when its feces were examined, the tortoise was found to be shedding oocysts of a *Cryptosporidium* sp., but the authors made no attempt to characterize it in any manner or to name it.

Cryptosporidium sp. of Xiao et al., 2004

Original host: *Chelonoidis carbonaria* (Spix, 1824) (syn. *Geochelone carbonaria* Ernst & Barbour, 1989), Red-footed Tortoise.

Remarks: Xiao et al. (2004) examined 123 fecal samples from captive snakes, lizards, and tortoises from the United States, Switzerland, the Czech Republic, Ghana, and Australia, and characterized the SSU rRNA gene of *Cryptosporidium*-positive samples by PCR-restriction fragment length polymorphism (PCR-RFLP) and DNA sequencing. Their results suggested extensive genetic diversity in these isolates from reptiles. They reported that 3/6 (50%) *Ch. carbonaria* from the Louisville Zoo, Kentucky, USA, were positive for *Cryptosporidium* by PCR, but these animals had not previously been screened for oocysts by light microscopy. They concluded that these tortoise isolates had slightly smaller *Ssp*I band size (near 800 bp), but had a *Vsp*I band similar to the band of *Cryptosporidium serpentis* (over 700 bp, from snakes). DNA sequencing showed that the sequences represented a new species of *Cryptosporidium* from this tortoise and a neighbor-joining analysis indicated that the tortoise genotype was related to *Cr. serpentis* and *Cryptosporidium muris*. However, they never named this new species.

Eimeria sp. of Bone, 1975 vide Wacha & Christiansen, 1976

Original host: *Graptemys barbouri* Carr & Marchand, 1942, Barbour's Map Turtle.

Remarks: Wacha and Christiansen (1976) found only one sporulated oocyst of an eimerian that they described, photographed, and provided a line drawing for, but did not name (p. 173). Following the description of their *Eimeria* sp., they mentioned that a colleague, L.W. Bone (University of Arkansas), had seen similar oocysts in *Gr. barbouri*, but that none of his oocysts ever developed beyond the four sporoblast stage. The partially sporulated oocysts measured by Bone ($N = 50$) were 18.6×14.4 ($18-21 \times 14-16$) with a L/W ratio of 1.3 (1.2–1.4). Unfortunately, Bone's description of a possible new species was never published.

Eimeria sp. of McAllister et al., 1994

Original host: *Glyptemys insculpta* (Le Conte, 1830) (syn. *Clemmys insculpta* Fitzinger, 1835), Wood Turtle.

Remarks: McAllister et al. (1994) found what they called, "An apparently undescribed *Eimeria* sp. in a captive *Gl. insculpta* from the Dallas (TX, USA) Zoo." Sporulated oocysts of this undescribed species were ellipsoidal and had an OR and a modest, but not large, SB. These latter features differ from those in oocysts of *Eimeria megalostiedai* described from a wood turtle in Iowa (Wacha & Christiansen, 1974). To our knowledge, a formal description of this eimerian was never published.

Eimeria sp. of Wacha & Christiansen, 1976

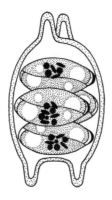

Figure 5.1 Line drawing of the sporulated oocyst of Eimeria *sp. from Wacha and Christiansen (1976), with permission from John Wiley & Sons, Ltd. holder of the copyright for the* Journal of Eukaryotic Microbiology *(formerly* Journal of Protozoology*).*

Original host: *Graptemys geographica* (Le Sueur, 1817), Common Map Turtle.

Remarks: Wacha and Christiansen (1976) found, measured, photographed (their fig. 15) and provided a line drawing (their fig. 9) for one sporulated oocyst, which was present in *Gr geographica* from which they examined the feces and the bile. The oocysts were narrowly ellipsoidal or truncated ovoidal, with two distinct points or conical projections at the slightly pointed (polar) end of the oocyst while the other end (anti-polar) is truncated into a flat base that is ornamented with two projections, giving the oocyst a mitra-like or projectile-like appearance. Oocyst characters such as M, PG, and OR were all absent. The single oocyst was 23.8×13.2 with a L/W ratio of 1.8. There were four narrowly ellipsoidal sporocysts that were all in a transverse position with their longitudinal axes parallel to each other. Each sporocyst had a small, thin, convex SB, but SSB and PSB were absent. The SR was a cluster of rod-shaped granules, each ~ 1. The only sporocyst they could measure was 11.2×5.3 and contained two SZ each with a RB at the broad end and a smaller RB at the narrow end. They found this oocyst in the feces and not in the bile, which led them to speculate it was an intestinal form. Interestingly, prior to publication (1976), they learned that a colleague at the University of Arkansas (Mr. L.W. Bone) had found an identically shaped, but somewhat larger *Eimeria* species in *Gr. barbouri* from Arkansas (see above for Bone's measurements).

Eimeria sp. 1 of Wacha & Christiansen, 1980

Type host: *Emydoidea blandingii* (Holbrook, 1838), Blanding's Turtle.

Remarks: In a survey of the coccidian parasites of Iowa turtles, two undescribed species of *Eimeria* were found. Those from Blanding's turtle were said to have ovoidal oocysts 25×13, with sporocysts 10×6; both OR and SR were present. To our knowledge, this species was never named or fully described. *Eimeria mitraria* also was found in this host species.

Eimeria sp. 2 of Wacha & Christiansen, 1980

Synonym: *Eimeria serpentina* McAllister et al., 1990.

Type host: *Chelydra serpentina* (L., 1758), Common Snapping Turtle.

Remarks: In a survey of Iowa turtles, two undescribed species of *Eimeria* were found. Those from *Ch. serpentina* had ovoidal to ellipsoidal oocysts, 15×9, that lacked an OR and had ovoidal sporocysts, 8.5×4, with an SR. To our knowledge, this species was never named or fully described by these authors. However, this unnamed eimerian from snappers in Iowa had oocysts that are similar to those of *E. serpentina* described and named by McAllister et al. (1990).

Mantonella hammondi Wacha & Christiansen, 1974b

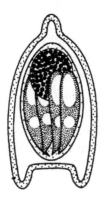

Figure 5.2 Line drawing of the sporulated oocyst of Mantonella hammondi *from Wacha and Christiansen (1976), with permission from John Wiley & Sons, Ltd. holder of the copyright for the* Journal of Eukaryotic Microbiology *(formerly* Journal of Protozoology*)*.

Original host: *Kinosternon flavescens spooneri* Wermuth & Mertens, 1977, Yellow Mud Turtle.

Remarks: The genus *Mantonella* Vincent, 1936 (syn. *Yakimovella* Gousseff, 1936) includes multiple species characterized by oocysts with one sporocyst, each containing four sporozoites. The three species from invertebrates appear valid, whereas the others most likely represent either pseudoparasites or are the result of improper sporulation conditions for the oocysts. Since crayfish comprise a portion of this turtle's diet, it seems likely that *M. hammondi* is a pseudoparasite. For the details of mensural data on oocysts, sporocysts, and SZ, see Wacha and Christiansen (1976, pp. 62–63).

Sarcocystis sp. of Keymer, 1978a

Original host: *Chelonoidis denticulatus* (L., 1766) (syn. *Testudo denticulata* L., 1766), Yellow-footed Tortoise.

Remarks: Keymer (1978a) mentioned finding *Sarcocystis*-like parasites in the skeletal muscle of the scapular region of a yellow-footed tortoise.

Sarcocystis sp. of Keymer, 1978b

Original host: *Kinosternon scorpioides* (L., 1766), Scorpion Mud Turtle.

Remarks: Keymer (1978b) mentioned finding *Sarcocystis*-like parasites in the skeletal muscle of a scorpion mud turtle.

Sarcocystis spp. of Meshkov, 1975

Original hosts: *Testudo graeca* L., 1758, Spur-thighed Tortoise and *Testudo hermanni* Gmelin, 1789, Hermann's Tortoise.

Remarks: Meshkov (1975) found microcysts of *Sarcocystis* in 43 of 49 (88%) tortoises of the above two species. The tortoises were collected from different regions of the two mountain massifs within the Bourgas District, the Balkan Range (six tortoises) and the Strandzha Mountain (37 tortoises), in Bulgaria. He did not mention the number of tortoises of each species collected, but he did say that all six (100%) from the Balkan Range were infected, while 31 (84%) from the Strandzha Mountain tortoises were infected. He noted that the microcysts were most often found in the muscles of the neck (81%) followed by the muscles of the hind leg (63%), and then the muscles of the front leg (53%). He concluded from his data that "sarcosporidia are not rare parasites in these animal species."

Sarcocystis sp. 1 of Weishaar et al., 1988

Original host: *Testudo horsfieldii* Gray, 1844, Horsfield's Tortoise.

Remarks: Weishaar et al. (1988) looked at European and South American tortoises as possible intermediate hosts for *Sarcocystis* species. They found *T. horsfieldii* to have tissue cysts in its skeletal muscles; the cysts had a smooth wall with compartments and could not be distinguished from cysts in other tortoise hosts they examined.

Sarcocystis sp. 2 of Weishaar et al., 1988

Original host: *Testudo marginata* Schoepff, 1792, Marginated Tortoise.

Remarks: Weishaar et al. (1988) looked at European and South American tortoises as possible intermediate hosts for *Sarcocystis* species. They found *T. marginata* to have tissue cysts in its skeletal muscles; the cysts had a smooth wall with compartments and could not be distinguished from cysts in other tortoise hosts they examined.

Sarcocystis sp. 3 of Weishaar et al., 1988

Original host: *Testudo graeca* L., 1758, Spur-thighed Tortoise.

Remarks: Weishaar et al. (1988) looked at European and South American tortoises as possible intermediate hosts for *Sarcocystis* species. They found *T. graeca* to have tissue cysts in its skeletal muscles; the cysts had a smooth wall with compartments and could not be distinguished from cysts in other tortoise hosts they examined.

Sarcocystis sp. 4 of Weishaar et al., 1988

Original host: *Testudo hermanni* Gmelin, 1789, Hermann's Tortoise.
Remarks: Weishaar et al. (1988) looked at European and South American tortoises as possible intermediate hosts for *Sarcocystis* species. They found *T. hermanni* to have tissue cysts in its skeletal muscles; the cysts had a smooth wall with compartments and could not be distinguished from cysts in other tortoise hosts they examined.

Sarcocystis sp. 5 of Weishaar et al., 1988

Original host: *Chelonoidis carbonaria* (Spix, 1824) (syn. *Testudo carbonaria* Spix, 1824), Red-footed Tortoise.
Remarks: Weishaar et al. (1988) looked at European and South American tortoises as possible intermediate hosts for *Sarcocystis* species. They found *Ch. carbonaria* (syn. *T. horsfieldii*) to have tissue cysts in its skeletal muscles; the cysts had a smooth wall with compartments and could not be distinguished from cysts in other tortoise hosts they examined.

Discussion and Summary

Order Testudines

BIODIVERSITY

Parasite biodiversity easily surpasses the diversity of their hosts (e.g., Windsor, 1988; Poulin & Morand, 2000), as we will demonstrate here. Although more than 120 years have passed since the first description of a coccidium in a turtle (Labbé, 1893), we still know precious little about how many coccidians (and other parasites) can and do infect turtles on Earth. As an example, Keymer (1978a,b) and Bourdeau (1989) listed the diseases and the etiological agents that caused them in their necropsies of more than 266 turtles and tortoises representing about 60 species. These disease agents included viruses and bacteria of both the skin and the intestinal tract; arthropods (e.g., dipterans), mites (e.g., trombiculids), and annelids on their external surfaces; blood (e.g., *Dracunculus* sp.) and intestinal (e.g., ascarids, pinworms) nematodes; and fungi, algae, protozoans (ciliates, amoebae, coccidia), acanthocephalans, pentastomids, and platyhelminthes (aspidogasters, monogeneans, trematodes, cestodes) of the gastrointestinal and respiratory tracts and other organs. Their surveys identified more than 150 organisms living in and on the turtles they examined. Ernst and Ernst (1979) compiled a synopsis of reported protozoans parasitic in native turtles in the United States at the time and listed almost 100 more protist species within various taxonomic categories. Such biodiversity represents just the tip of the iceberg of what we might find in turtles if all their species could be thoroughly sampled.

Within just the coccidian parasites of vertebrates, the vast majority of taxa still await discovery. For example, in the Amphibia, only 45 of 6009 known species (<1%) have been examined for coccidia, but about 87 species/forms have been described (~52 are considered valid species) (Duszynski et al., 2007); in the Serpentes, only 208 of 3180 known species (6.5%) have been examined for coccidia, but about 303 species/forms have been described (~155 are valid) (Duszynski & Upton, 2009); in the

The Biology and Identification of the Coccidia (Apicomplexa) of Turtles of the World.
DOI: http://dx.doi.org/10.1016/B978-0-12-801367-0.00006-X

Chiroptera (bats), only 86 of 925 known species (9%) have been examined for coccidia, but 40 species/forms have been named and described (~32 are valid) (Duszynski, 2002); in the Insectivora (moles, shrews) only 37 of the 428 known species (8.6%) have been examined for coccidia, but 120 species/forms have been named and described (~75 are valid) (Duszynski & Upton, 2000); within the Scandentia (tree shrews), only four of the 19 known species (21%) have been examined for coccidia, and four *Eimeria* have been named (Duszynski et al., 1999); within the Primates (Prosimii, Anthropoidea) only 18 of the 233 known species (8%) have been examined for coccidia (Duszynski et al., 1999) and nine species have been named; and within the Lagomorpha (rabbits, hares) only 23 of the 91 known species (25%) have been examined for coccidia, but 114 species/forms have been named and described (~87 are valid) (Duszynski & Couch, 2013). These studies illustrate that the vast majority of our knowledge about the coccidia of wild vertebrates has focused on mammals. We also know a great deal about the coccidia of some birds (e.g., domestic chickens) because of their importance as a food source for humans but, as noted above, little is known about the coccidia and other parasites from reptiles. The monoxenous coccidia make excellent organisms for the study of biodiversity because they are obtained from the feces of their hosts by noninvasive methods and their exogenous oocysts have relatively simple morphology, which can be studied easily with the light microscope.

Increases in intensive pet trade of amphibians and reptiles worldwide have led to the spread of numerous vertebrate species, including turtles, around the globe. Since their associated parasites also can be distributed with their hosts, our understanding of the role(s) of geographic origin and isolation of both partners, the host and the parasite, has been greatly diminished. Thus, some authors have decided to name coccidia with sporulated oocysts that are morphologically similar and found in related genera of the same chelonid family, as the same species (e.g., Široký & Modrý, 2010). But are they correct in doing so?

VARIETY OF OOCYST STRUCTURES AND SHAPE

When we look at the shape and structure of the oocyst walls of species of *Caryospora*, *Eimeria*, and *Isospora* reported from turtles we see unbridled diversity (line drawings, Chapters 2 and 3). Why? What is the evolutionary significance of this variation? Why do some chelonid coccidia have oocysts with a micropyle while others do not? Why do some have tubercles on their walls while most don't? What is the significance of having an oocyst residuum and a polar body while some have one or the other, and yet others have neither (see Table 4)?

Only five of 70 (7%) coccidians in the three major genera have oocysts that exhibit a micropyle and three of these five have oocysts with very thin (≤ 1.0) walls that are easily ruptured. So it is a mystery why a micropyle could or should arise. As first mentioned by Lainson et al. (1990), until we understand the complete life history of all chelonid coccidia, we can only speculate that, perhaps, some peculiarities in the transfer of these parasites to new, susceptible, hosts require the presence of a micropyle to more readily liberate their sporozoites.

None of the oocysts of *Caryospora* and *Isospora* species known to date from turtles have an oocyst residuum, but half, 33/66 (50%), of the eimerian sporulated oocysts identified from chelonids have one. It has been shown in previous studies on rodents and bats that this morphological feature shows a clear correlation to the phylogenetic relationship as demonstrated by molecular genetic methods (Zhao & Duszynski, 2001a,b; Zhao et al., 2001). There have been no molecular studies done on oocysts from turtle species, but Široký and Modrý (2006) indicated the probability that future molecular work on these species may reveal phylogenies showing even more such "subgroups" within this genus.

One morphological feature exhibited by the vast majority (57/70, 81%) of the oocysts of turtle coccidians is that they have very thin (≤ 1.0) and fragile walls, with the remainder having walls that are only slightly thicker ($1.1-1.5$) (Table 4). This seems a logical outcome for hosts that are usually associated with an aquatic environment because thick outer walls would be highly energetic to produce when it is not necessary to protect the contents of the oocyst from desiccation.

One of the striking features of the oocyst walls of 9/70 (13%) of the coccidians described from turtles is the presence of $2-11$ peculiar projections of their oocyst walls. It is also of interest that in every instance, these oocyts lack a micropyle, oocyst residuum, and a polar granule, and all of them have very thin walls. Široký and Modrý (2006) suggested that species with such oocysts were probably a case of synapomorphy rather than a convergence, and actually represent a separate evolutionary lineage, the emergence of which predates the separation of continents. One species that exhibits these peculiar tubercles on the oocyst wall, *E. mitraria*, has been reported from 15 turtle species or subspecies in both extant suborders (Table 1) and on distant continents. Thus, we are inclined to agree with Široký and Modrý (2006) that as presently used, *E. mitraria* may represent "a 'morphotype' rather than a species."

HOST SPECIFICITY

We know almost nothing about host specificity in turtles because only one cross-transmission event has been done experimentally. Carini (1942) said that he attempted to infect other *Chelonoidis* spp. (he called them *Testudo* spp.) adults by making them eat feces containing mature oocysts of *E. jaboti*, but that his attempts were all negative. Additionally, we don't know how natural transmission of oocysts might occur within and/or between species in any particular habitat.

Široký and Modrý (2005, 2006, 2010), Široký et al. (2006a,b), and McAllister and Upton (1988, 1989a,b, 1990) have surveyed turtles from many families around the world. If we look at their work as a whole, some of which included a careful analysis of the literature, it may suggest that most species of coccidia in the Testudines are specific at the family level. For example, painted turtles (*Chrysemys*), cooters (*Pseudemys*), map turtles (*Graptemys*), and sliders (*Trachemys*), all in the family Emydidae, seem to share several *Eimeria* species (e.g., see McAllister & Upton, 1988, 1989b); similarly, yellow-margined box turtles (*Cuora*) and Arakan forest turtles (*Heosemys*), both in the family Geomydidae, also share some species. Other eimeriids, based on the limited surveys available, may be restricted only to certain turtle genera, however, this cannot be confirmed or refuted based on available evidence. Examples of possible generic restrictions might include *E. trachemydis* in *T. s. elegans*, several eimeriid species in *Pseudemys texana*, as well as *E. carri* and *E. ornata* in box turtles, *Terrapene* spp. (McAllister & Upton, 1988, 1989a,b, 1990).

At the other extreme, at least one species, *E. mitraria*, has been recorded from 15 turtle species or subspecies representing 10 genera in four families (Chelydridae, Emydidae, Geomydidae, and Kinosternidae; see Table 1 in Široký & Modrý, 2006). However, McAllister et al. (1990a) argued that their work did not support *C. serpentina* as a valid host for *E. mitraria* because even when this parasite was prevalent in Emydidae from a single locale, snapping turtles collected from the same pond were not found to harbor this species (McAllister et al., 1990a,b). They argued further that a morphologically similar coccidium, *I. chelydrae*, occurs in *C. serpentina* and that this may have been the coccidian seen by Wacha and Christiansen (1980). Their final explanation of why *E. mitraria* may not be a real parasite of *C. serpentina* is that it may represent a pseudoparasite. That is, snapping turtles are omnivorous and will eat anything available, including other turtles (Collins, 1982). Thus, "Considering the high prevalences of *E. mitraria* in other sympatric turtle species, oocysts would probably be detected periodically in the feces of *C. serpentina*" (McAllister et al., 1990a).

Lainson and Naiff (1998) quoted McAllister and Upton (1989) and McAllister et al. (1990) saying, "...most, but not all, of the turtle Coccidia from aqueous environments in North America are not particularly species specific," and "...most species of coccidia in the Testudines are specific at the family level" (McAllister et al., 1990). Lainson and Naiff (1998) went on to suggest that the literature and data they assembled (their Table I) supported this view, with some *Eimeria* species (*E. marginata*, *E. tetradacrutata*) recorded in three, four (*E. graptemydos*, *E. lutotestudinis*), or even an astonishing eight different genera of chelonians (*E. mitraria*). Building largely from data collected in North America, they concluded: "Much less is known about the host range and prevalence of the coccidia of chelonians in the neotropics and the Old World. It would be strange, however, if a similar situation does not exist in these regions." In evaluating all the literature to date regarding turtle coccidia, the degree of host specificity within these host–parasite relationships remains elusive. This ambiguity is largely a factor of sampling bias coupled with a lack of systematic sampling over a broad host geographic range; that is, sampling of turtles for coccidia has been restricted to specific regions, each with their own unique turtle population, sampled by a small set of interested researchers (see Table 3). Perhaps with comprehensive area and specimen sampling in the future, a clearer understanding of host specificity among these parasites will come to fruition.

PATHOLOGY

Most species of coccidia that develop in turtles are "considered" to be nonpathogenic, but this may reflect methodologies more so than reality. Most often, researchers sample feces to report the presence/absence of these parasites and are not looking for, or are not able to look for, a "disease" condition or symptoms. For example, many/most hosts collected in the field look healthy even though they are infected with intracellular parasites that kill many of their cells, often epithelial cells in the digestive system and related organ systems (e.g., bile duct). Evidently, the cell turnover rate is such in these locations, and the destroyed cells are replaced quickly enough, that the host can remain "healthy" in appearance. When some coccidia do manifest a disease condition in their host, it is usually associated with a high dose of infective oocysts that are ingested over a short time period. This fact may be a reason why many coccidian infections in turtles (and other reptiles) are not significant to host health in nature. Since the route of infection, in most cases, is fecal–oral, the volume of contaminated material ingested by most reptiles would be expected to be small (Greiner, 2003).

In the die-off of green turtles recorded in the British West Indies (Rebell et al., 1974; Leibovitz et al., 1978), and discussed earlier in detail, it was noted that food and fecal sludge had been permitted to accumulate in the bottom of the holding tanks in which young turtles were placed after hatching. Upon inspection, large numbers of unsporulated oocysts were found in this anaerobic bottom sludge. Spread of the epidemic in a row of tanks occurred rapidly, possibly as a result of transferring turtles sequentially from tank to tank during cleaning by using a holding tank. Several hundred turtle hatchlings that were raised separately and not placed in the holding tanks failed to develop symptoms of coccidiosis and remained uninfected. This study implicated and highlighted the importance of management practices in the rearing of turtles.

EPIDEMIOLOGY

Work by Široký and Modrý (2005, 2006, 2010) and Široký et al. (2006a,b) suggests that the Asian zoogeographic region may well represent a hot spot for turtle eimeriid biodiversity because they were able to describe one, two, or more new species from virtually every host they examined. Many of the cheloinian species are listed by the IUCN as endangered or critically endangered (IUCN, 2009) and many of these species will become extinct long before they can be examined for their parasites. Thus, the unexplored biodiversity of their parasites can potentially be lost forever following the collapse of the natural populations of these turtle hosts as they are incessantly hunted as a food source, as a source for regional, traditional medicines, as well as for their export in the ever-expanding pet trade. There are tremendous numbers of these protist parasites yet to be discovered in turtles, but these myriads of parasites will be lost forever as their hosts continue to disappear from Earth through loss of habitat, toxins in the environment, climate change, and the untold other environmental perturbations of human origin. Yes, even many parasite species are endangered!

TREATMENT AND CONTROL

The occurrence of coccidiosis among wild turtle populations infected with *Eimeria* and/or *Isospora* species seems unlikely as these animals will naturally ingest only occasional sporulated oocysts from their unregulated environments. Thus, treatment of wild populations will most often be unnecessary as infections will run their course in immunocompetent individuals and continuous exposure to repeated light infections should stimulate absolute immunity. In captive turtle populations, however, some coccidians (e.g., *Caryospora*) and other protists (e.g., *Cryptosporidium*) can cause serious problems if measures are not taken to prevent transmission.

During the 1973 outbreak of coccidiosis due to *Caryospora cheloniae*, controlled experiments were performed with small groups of 25–100 turtles taken from the tanks in which the infections were first detected to test the effect of tetracycline and sulfamethazine (an anticoccidial sulfonamide) that was administered by ingestion, intraperitoneal injection, and overnight soaks (Rebell et al., 1974). No positive effects were obtained from these experiments, but this treatment regime was not attempted until late in the infection epidemic. In sick animals, no significant differences were noted between the mortality of treated and untreated groups. Likely, this was because the parasite had already run the course of its endogenous development.

Pavlásek (1998) examined an Amboina box turtle, *C. amboinensis*, which came from Vietnam. The turtle had watery diarrhea with numerous oocysts of a *Cryptosporidium* species, which he attributed to the pathology observed. Captive turtles and tortoises that become exposed to coccidia and/or *Cryptosporidium* species following poor preventative measures can be treated in a manner similar to the situation in humans with HIV because in both cases the disease is not self-limiting (Cranfield & Graczyk, 2000). For reptiles, a regimen of tri-methoprim—sulfamethoxazole (TMP–SMZ, Biocraft Laboratories), 30 mg/kg once a day for 14 days and then one to three times weekly for several months, and spiramycin, 160 mg/kg for 10 days, and paromomycin, 100 mg/kg for 7 days, and then twice a week for 3 months, seem to be effective for reducing the magnitude of clinical signs of disease and decreasing or eliminating the shedding of oocysts (Cranfield & Graczyk, 2000). The recommended treatment regimens are identical to those recommended for treating snake coccidiosis. Other recommended regimens can be found in Duszynski and Upton (2009, table 12, pp. 312–316). These authors, along with Graczyk et al. (1996, 1998), also outlined prevention strategies for snakes that may be applicable to the prevention of turtle coccidiosis.

ARCHIVING BIOLOGICAL SPECIMENS

As noted in Chapter 2, we strongly encourage both parasitologists and herpetologists who do field collections, to always archive specimens of both coccidian parasites and their hosts in accredited museums from which other scientists can access them. Parasitologists should save DNA, either fixed or with sequences stored in GenBank. Saving the DNA is a simple matter of storing oocysts in 70–100% ethanol (EtOH), if the person collecting this material is unable to sequence genes at the time collections of oocysts are made. Arthropod and vertebrate biologists are far ahead of us on this front because they have been preparing their collected materials as proper study skins (mammals, birds) or immersed in 70% EtOH or 10% formal saline (amphibians, fishes, arthropods, reptiles) for decades so that future workers can use them to amplify DNA as our techniques and protocols improve. Other methods to preserve coccidia specimens are histological sections of infected tissues on slides, or tissues fixed and embedded in paraffin or plastic to be sectioned in the future. Photosyntypes (digitized photomicrographs) of sporulated oocysts should always be archived in an accredited museum so other scientists can access to them. To date, of the 70 valid coccidia species named in this book (excluding *Cryptosporidium* and *Sarcocystis*), either host or parasite "type" materials, or both, have been archived for 41 (59%) in a museum. The symbiotype host (Frey et al., 1992) alone has been archived for three species (4%); the symbiotype host and oocysts in formalin have been archived for five species (7%); the symbiotype host and photosyntype (Duszynski, 1999) images of sporulated oocysts have been archived for three species (4%); oocysts in formalin have been archived for eight species (11%); and photosyntypes of sporulated oocysts alone have been archived for 20 species (28.5%). This is a good start, but we can, and should, continue to archive all protist materials to improve specimen collections in the future for both host and new protist species.

CLOSING REMARKS

The information presented here has been compiled to aid future herpetologists, and the parasitologists who work with them, to better understand the role of coccidia and related protists in the biology of each turtle species. We hope future scholars interested in the interface between turtles and their coccidia can use this book as a guide to easily assess the voids in our knowledge regarding these host–parasite systems. Knowing what has been accomplished, and what is known, should be useful to more easily identify areas that still need to be explored. Developing research projects in the areas that have not yet been studied will ultimately provide meaningful contributions to the field of both turtle biology and coccidian research. Only from careful examination and analysis of the existing body of literature can valuable new initiatives in this research area be initiated.

There is more that we don't know about turtle coccidia than what we do know, *much more*. To wit, most questions still remain unanswered on the general biology, taxonomy, distribution, epidemiology, ecology, life cycles, endogenous development, pathology, host specificity, and other areas. The inherent sampling bias (see Table 3) of our current data only can be more meaningful with systematic and thorough sampling of all turtle species and populations worldwide, meticulous documentation of all findings, a strong laboratory component to help document life history and molecular data, and diligent reporting of those findings in peer-reviewed scientific journals. We strongly encourage anyone interested in turtles and their biology to consider our work as a starting point to help them formulate meaningful conceptual research questions to begin their research in the area of turtle biology and their parasites.

TABLES

Table 1 Alphabetical List of All Coccidian Parasites Covered in this Book and the Turtle Hosts from Which They Have Been Reported

Caryospora cheloniae
 Chelonia mydas mydas
***Caryospora* sp.** (species inquirenda)
 Astrochelys radiate
***Coccidium* spp.** (species inquirendae)
 Astrochelys radiata
 Chersina angulata
 Indotestudo forstenii
 Manouria impressa
 Stigmochelys pardalis
***Cryptosporidium* spp.** (species inquirendae)
 Apalone spinifera hartwegi
 Chelonia mydas
 Chelonoidis carbonaria
 Geochelone elegans
 Glyptemys muhlenbergi
 Kinixys sp.
 Testudo kleinmanni
Eimeria amazonensis
 Chelonoidis carbonaria
Eimeria amydae
 Apalone spinifera
 Apalone spinifera pallida
Eimeria apalone
 Apalone spinifera pallida
 Apalone spinifera hartwegi
Eimeria arakanensis
 Heosemys depressa
 Cuora flavomarginata
Eimeria brodeni
 Testudo graeca
Eimeria carajasensis
 Chelonoidis carbonaria
Eimeria carbonaria
 Chelonoidis carbonaria
Eimeria caretta
 Caretta caretta
Eimeria carri
 Terrapene carolina
 Terrapene carolina carolina
 Terrapene carolina triunguis
 Terrapene ornata ornata
Eimeria chelydrae
 Chelydra serpentine
Eimeria chrysemydis
 Chrysemys picta belli
 Graptemys caglei

(Continued)

Table 1 (Continued)

    Graptemys geographica
    Trachemys gaigeae
    Trachemys scripta elegans
Eimeria cooteri
    Pseudemys texana
Eimeria delagei
    Emys orbicularis
Eimeria dericksoni
    Apalone spinifera
    Apalone spinifera hartwegi
    Apalone spinifera pallida
Eimeria emydis
    Emys orbicularis
Eimeria filamentifera
    Chelydra serpentina
Eimeria gallaeciaensis
    Emys orbicularis
Eimeria geochelona
    Chelonoidis nigra
Eimeria graptemydos
    Chrysemys picta belli
    Chrysemys picta dorsalis
    Graptemys caglei
    Graptemys geographica
    Graptemys versa
    Kinosternon flavescens flavescens
    Kinosternon subrubrum hippocrepis
    Trachemys gaigeae
    Trachemys scripta elegans
Eimeria harlani
    Macrochelys temminckii
Eimeria hynekprokopi
    Cuora galbinifrons
Eimeria innominata
    Lissemys punctata
Eimeria irregularis
    Lissemys punctata
Eimeria iversoni
    Chelonoidis sp.
Eimeria jaboti
    Chelonoidis denticulata
Eimeria jirkamoraveci
    Mesoclemmys heliostemma
Eimeria juniataensis
    Graptemys geographica
Eimeria kachua
    Pangshura tentoria circumdata
Eimeria koormae
    Lissemys punctata
Eimeria lagunculata
    Podocnemis expansa
Eimeria lainsoni
    Chelonoidis denticulata
Eimeria lecontei
    Glyptemys insculpta
Eimeria légeri
    Lissemys punctata

(Continued)

Table 1 (Continued)

Eimeria lokuma
    Pelomedusa subrufa
Eimeria lutotestudinis
    Graptemys caglei
    Graptemys geographica
    Kinosternon flavescens flavescens
    Kinosternon flavescens spooneri
    Kinosternon subrubrum hippocrepis
    Pseudemys texana
    Trachemys scripta elegans
Eimeria mammiformis
    Podocnemis expansa
Eimeria marginata
    Chrysemys picta belli
    Graptemys geographica
    Graptemys pseudogeographica
    Pseudemys concinna
    Trachemys gaigeae
    Trachemys scripta elegans
Eimeria mascoutini
    Apalone spinifera
    Apalone spinifera pallida
Eimeria megalostiedae
    Glyptemys insculpta
Eimeria mitraria
    Chelydra serpentina serpentina
    Chrysemys picta belli
    Emydoidea blandingii
    Emys orbicularis
    Graptemys geographica
    Graptemys pseudogeographica
    Graptemys versa
    Heosemys depressa
    Kinosternon flavescens flavescens
    Kinosternon flavescens spooneri
    Mauremys reevesii
    Mesoclemmys heliostemma
    Pseudemys texana
    Terrapene carolina triunguis
    Trachemys scripta elegans
Eimeria motelo
    Chelonoidis denticulate
Eimeria ornata
    Terrapene ornata ornata
Eimeria palawanensis
    Cyclemys dentata
Eimeria pallidus
    Apalone spinifera pallida
Eimeria pangshurae
    Pangshura sylhetensis
Eimeria patta
    Melanochelys trijuga
Eimeria paynei
    Gopherus polyphemus
Eimeria peltocephali
    Peltocephalus dumerilianus

(Continued)

Table 1 (Continued)

Eimeria petrasi
    Cyclemys dentata
Eimeria podocnemis
    Podocnemis expansa
Eimeria pseudemydis
    Deirochelys reticularia miaria
    Glyptemys insculpta
    Pseudemys texana
    Trachemys gaigeae
    Trachemys ornata
    Trachemys scripta elegans
Eimeria pseudogeographica
    Chrysemys picta belli
    Graptemys caglei
    Graptemys ouachitensis
    Graptemys pseudogeographica
    Trachemys gaigeae
    Trachemys scripta elegans
Eimeria scriptae
    Trachemys scripta elegans
Eimeria serpentina
    Chelydra serpentina serpentina
Eimeria somervellensis
    Pseudemys concinna metteri
    Pseudemys texana
Eimeria spinifera
    Apalone spinifera pallida
Eimeria **spp.** (species inquirendae)
    Chelonoidis denticulata
    Chelydra serpentina
    Emydoidea blandingii
    Glyptemys insculpta
    Graptemys barbouri
    Graptemys geographica
    Heosemys spinosa
    Pseudemys texana
    Trachemys gaigeae
Eimeria stylosa
    Trachemys scripta elegans
    Trachemys gaigeae
Eimeria surinensis
    Malayemys subtrijuga
Eimeria tetradacrutata
    Chrysemys picta belli
    Graptemys geographica
    Trachemys scripta elegans
Eimeria texana
    Pseudemys texana
Eimeria trachemydis
    Chrysemys picta belli
    Graptemys caglei
    Trachemys gaigeae
    Trachemys scripta elegans
Eimeria triangularis
    Nilssonia gangetica
Eimeria trionyxae
    Nilssonia gangetica

(*Continued*)

Table 1 (Continued)

Eimeria vesicostieda
    Apalone spinifera
    Apalone spinifera hartwegi
Eimeria welcomei
    Chelonoidis carbonaria
Eimeria zbatagura
    Batagur baska
Isospora chelydrae
    Chelydra serpentina serpentina
Isospora rodriguesae
    Chelonoidis denticulata
Isospora testudae
    Testudo horsfieldii
Mantonella hammondi (species inquirenda)
    Kinosternon flavescens spooneri
Sarcocystis kinosterni
    Kinosternon scorpioides
***Sarcocystis* spp.** (species inquirendae)
    Chelonoidis carbonaria
    Chelonoidis denticulata
    Kinosternon scorpiodes
    Testudo graeca
    Testudo hermanni
    Testudo horsfieldii
    Testudo marginata

Table 2 Alphabetical List of All Turtle Species Covered in This Book and the Coccidian Parasites Which Have Been Reported in Them

Suborder Cryptodira, Hidden-Necked Turtles

Apalone spinifera
    Eimeria amydae
    Eimeria dericksoni
    Eimeria mascoutini
    Eimeria vesicostieda
Apalone spinifera hartwegi
    Eimeria dericksoni
    Eimeria vesicostieda
    Cryptosporidium sp. (species inquirenda)
Apalone spinifera pallida
    Eimeria amydae
    Eimeria apalone
    Eimeria dericksoni
    Eimeria mascoutini
    Eimeria pallidus
    Eimeria spinifera
Astrochelys radiata
    Coccidium sp. (species inquirenda)
    Caryospora sp. (species inquirenda)
Batagur baska
    Eimeria zbatagura
Caretta caretta
    Eimeria caretta

(Continued)

Table 2 (Continued)

Chelonia mydas
    Cryptosporidium sp. (species inquirenda)
Chelonia mydas mydas
    Caryospora cheloniae
Chelonoidis carbonaria
    Eimeria amazonensis
    Cryptosporidium sp. (species inquirenda)
    Eimeria carajasensis
    Eimeria carbonaria
    Eimeria welcomei
    Sarcocystis sp. (species inquirenda)
Chelonoidis denticulata
    Eimeria jaboti
    Eimeria lainsoni
    Eimeria motelo
    Eimeria sp. (species inquirenda)
    Isospora rodriguesae
    Sarcocystis sp. (species inquirenda)
Chelonoidis nigra
    Eimeria geochelona
Chelonoidis sp.
    Eimeria iversoni
Chelydra serpentina
    Eimeria chelydrae
    Eimeria filamentifera
    Eimeria sp. (species inquirenda)
Chelydra serpentina serpentina
    Eimeria mitraria
    Eimeria serpentina
    Isospora chelydrae
Chersina angulata
    Coccidium sp. (species inquirenda)
Chrysemys picta belli
    Eimeria chrysemydis
    Eimeria graptemydos
    Eimeria marginata
    Eimeria mitraria
    Eimeria pseudogeographica
    Eimeria tetradacrutata
    Eimeria trachemydis
Chrysemys picta dorsalis
    Eimeria graptemydos
Cuora flavomarginata
    Eimeria arakanensis
Cuora galbinifrons
    Eimeria hynekprokopi
Cyclemys dentata
    Eimeria palawanensis
    Eimeria petrasi
Deirochelys reticularia miaria
    Eimeria pseudemydis
Emydoidea blandingii
    Eimeria mitraria
    Eimeria sp. (species inquirenda)
Emys orbicularis
    Eimeria delagei
    Eimeria emydis

(Continued)

Table 2 (Continued)

  Eimeria gallaeciaensis
  Eimeria mitraria
Geochelone elegans
  Cryptosporidium sp. (species inquirenda)
Glyptemys insculpta
  Eimeria lecontei
  Eimeria megalostiedae
  Eimeria pseudemydis
  Eimeria sp. (species inquirenda)
Glyptemys muhlenbergi
  Cryptosporidium sp. (species inquirenda)
Gopherus polyphemus
  Eimeria paynei
Graptemys barbouri
  Eimeria sp. (species inquirenda)
Graptemys caglei
  Eimeria chrysemydis
  Eimeria graptemydos
  Eimeria lutotestudinis
  Eimeria pseudogeographica
  Eimeria trachemydis
Graptemys geographica
  Eimeria chrysemydis
  Eimeria graptemydos
  Eimeria juniataensis
  Eimeria lutotestudinis
  Eimeria marginata
  Eimeria mitraria
  Eimeria sp. (species inquirenda)
  Eimeria tetradacrutata
Graptemys ouachitensis
  Eimeria pseudogeographica
Graptemys pseudogeographica
  Eimeria marginata
  Eimeria mitraria
  Eimeria pseudogeographica
Graptemys versa
  Eimeria graptemydos
  Eimeria mitraria
Heosemys depressa
  Eimeria arakanensis
  Eimeria mitraria
Heosemys spinosa
  Eimeria sp. (species inquirenda)
Indotestudo forstenii
  Coccidium sp. (species inquirenda)
Kinixys sp.
  Cryptosporidium sp. (species inquirenda)
Kinosternon flavescens flavescens
  Eimeria graptemydos
  Eimeria lutotestudinis
  Eimeria mitraria
Kinosternon flavescens spooneri
  Eimeria lutotestudinis
  Eimeria mitraria
  Mantonella hammondi (species inquirenda)

(Continued)

Table 2 (Continued)

Kinosternon scorpioides
    Sarcocystis kinosterni
    Sarcocystis sp. (species inquirenda)
Kinosternon subrubrum hippocrepis
    Eimeria graptemydos
    Eimeria lutotestudinis
Lissemys punctata
    Eimeria innominata
    Eimeria irregularis
    Eimeria koormae
    Eimeria légeri
Macrochelys temminckii
    Eimeria harlani
Malayemys subtrijuga
    Eimeria surinensis
Manouria impressa
    Coccidium sp. (species inquirenda)
Mauremys reevesii
    Eimeria mitraria
Melanochelys trijuga
    Eimeria patta
Mesoclemmys heliostemma
    Eimeria mitraria
Nilssonia gangetica
    Eimeria triangularis
    Eimeria trionyxae
Pangshura sylhetensis
    Eimeria pangshurae
Pangshura tentoria circumdata
    Eimeria kachua
Pseudemys concinna
    Eimeria marginata
Pseudemys concinna metteri
    Eimeria somervellensis
Pseudemys texana
    Eimeria cooteri
    Eimeria lutotestudinis
    Eimeria mitraria
    Eimeria pseudemydis
    Eimeria sp. (species inquirenda)
    Eimeria somervellensis
    Eimeria texana
Stigmochelys pardalis
    Coccidium sp. (species inquirenda)
Terrapene carolina
    Eimeria carri
Terrapene carolina carolina
    Eimeria carri
Terrapene carolina triunguis
    Eimeria carri
    Eimeria mitraria
Terrapene ornata ornata
    Eimeria carri
    Eimeria ornata
Testudo graeca
    Eimeria brodeni
    Sarcocystis sp. (species inquirenda)

(Continued)

Table 2 (Continued)

Testudo hermanni
 Sarcocystis sp. (species inquirenda)
Testudo horsfieldii
 Isospora testudae
 Sarcocystis sp. (species inquirenda)
Testudo kleinmanni
 Cryptosporidium sp. (species inquirenda)
Testudo marginata
 Sarcocystis sp. (species inquirenda)
Trachemys scripta elegans
 Eimeria chrysemydis
 Eimeria graptemydos
 Eimeria lutotestudinis
 Eimeria marginata
 Eimeria mitraria
 Eimeria pseudemydis
 Eimeria pseudogeographica
 Eimeria scriptae
 Eimeria stylosa
 Eimeria tetradacrutata
 Eimeria trachemydis
Trachemys gaigeae
 Eimeria chrysemydis
 Eimeria graptemydos
 Eimeria marginata
 Eimeria pseudemydis
 Eimeria pseudogeographica
 Eimeria sp. (species inquirenda)
 Eimeria stylosa
 Eimeria trachemydis
Trachemys ornata
 Eimeria pseudemydis

Suborder Pleurodira, Side-Necked Turtles

Mesoclemmys heliostemma
 Eimeria jirkamoraveci
 Eimeria mitraria
Pelomedusa subrufa
 Eimeria lokuma
Peltocephalus dumerilianus
 Eimeria peltocephali
Podocnemis expansa
 Eimeria lagunculata
 Eimeria mammiformis
 Eimeria podocnemis

Table 3 Alphabetical List of All the Continents, Countries, and States from Which Turtle Coccidian Species Covered in This Book were Originally Described (Type Locality)

ASIA (10 Countries, 17 Species)

Bangladesh—*Eimeria koormae*
India—*E. innominata, E. irregularis, E. kachua, E. légeri, E. pangshurae, E. triangularis, E. trionyxae*
Japan (?)—*E. mitraria*
Myanmar (Burma)—*E. patta*
Philippines—*E. palawanensis, E. petrasi*
Singapore—*E. zbatagura*
Thailand—*E. surinensis*
Uzbekistan—*Isospora testudae*
Vietnam—*E. hynekprokopi*
Western Myanmar—*E. arakanensis*

CARIBBEAN (1 Species)

British West Indies—*Caryospora cheloniae*

CENTRAL AMERICA (1 Species)

Belize—*E. pseudemydis*

NORTH AMERICA: USA (8 States, 31 Species)

Alabama—*E. carri*
Arkansas—*E. harlani, E. serpentina, I. chelydrae*
Florida—*E. caretta*
Georgia—*E. chelydrae, E. paynei*
Iowa—*E. amydae, E. chrysemydis, E. dericksoni, E. graptemydos, E. lutotestudinis, E. marginata, E. mascoutini, E. megalostiedae, E. pseudogeographica, E. tetradacrutata, E. vesticostieda*
Pennsylvania—*E. juniataensis*
Texas—*E. apalone, E. cooteri, E. iversoni, E. lecontei, E. ornata, E. pallidus, E. somervellensis, E. spinifera, E. stylosa, E. texana, E. trachemydis*
Wisconsin—*Eimeria scriptae*

SOUTH AMERICA (3 Countries, 7 Species)

Brazil—*E. amazonensis, E. carajasensis, E. carbonaria, E. jaboti, E. lainsoni, E.welcomei, I. rodriguesae*
Ecuador—*E. geochelona*
Peru—*E. motelo*

WESTERN EUROPE (3 Countries, 4 Species)

France—*E. delagei*
Italy—*E. brodeni*
Spain—*E. emydis, E. gallaeciaensis*

Table 4 Morphological Characters of the Oocysts of All *Eimeria, Caryospora,* and *Isospora* Species Described in Chapters 2 and 3, in the Order in Which They Appear in These Chapters

Turtle Suborder/Coccidia Species	Oocyst Structures[a]			Wall Thickness		Wall with Tubercles/ Points
	M	OR	PG	≤1.0	≥1.1	
Cryptodira						
Eimeria chelydrae	−	−	−	+		−
Eimeria filamentifera	−	+	−	+		−
Eimeria serpentina	−	−	−	+		−
Eimeria harlani	−	+	+	+		−
Eimeria chrysemydis	−	+	±	+		−
Eimeria marginata	−	+	+	+		−
Eimeria tetradacrutata	−	+	+		+	−
Eimeria delagei	−	+	−	+		−
Eimeria emydis	−	+	−	+		−
Eimeria gallaeciaensis	−	+	−	+		−
Eimeria lecontei	−	+	+	+		−
Eimeria megalostiedae	−	+	−	+		−
Eimeria graptemydos	−	+	+	+		−
Eimeria juniataensis	−	+	−	+		−
Eimeria pseudogeographica	−	+	+	+		−
Eimeria cooteri	−	+	+	+		−
Eimeria somervellensis	−	+	+	+		−
Eimeria texana	−	+	+	+		−
Eimeria carri	−	+	−	+		−
Eimeria ornata	−	+	±	+		−
Eimeria pseudemydis	−	+	±		+	−
Eimeria scriptae	+	−	−		+	−
Eimeria stylosa	−	−	−	+		4–11
Eimeria trachemydis	−	+	+	+		−
Eimeria amazonensis	−	−	−	+		2
Eimeria carajasensis	−	−	+		+	−
Eimeria carbonaria	−	−	+	+		−
Eimeria geochelona	−	−	+		+	−
Eimeria iversoni	−	−	−	+		2
Eimeria jaboti	−	−	+		?	−

(Continued)

Table 4 (Continued)

Turtle Suborder/Coccidia	Oocyst Structures[a]			Wall Thickness		Wall with Tubercles/
Species	M	OR	PG	≤1.0	≥1.1	Points
Eimeria lainsoni	−	+	−		+	−
Eimeria motelo	−	−	−	+		2
Eimeria welcomei	−	−	−	+		−
Eimeria paynei	−	−	+		+	−
Eimeria brodeni	+	−	−	?		−
Eimeria zbatagura	−	−	−	+		−
Eimeria hynekprokopi	−	−	−	+		2−4
Eimeria palawanensis	−	+	−	+		−
Eimeria petrasi	−	+	+	+		−
Eimeria arakanensis	−	+	−	+		−
Eimeria surinensis	−	+	−		+	−
Eimeria mitraria	−	−	−	+		4−5
Eimeria patta	−	−	−	+		−
Eimeria pangshurae	−	−	+	+		−
Eimeria kachua	−	+	+	+		−
Eimeria amydae	−	+	−	+		−
Eimeria apalone	−	−	−	+		−
Eimeria dickersoni	−	+	±	+		−
Eimeria mascoutini	−	−	+	+		−
Eimeria pallidus	−	+	±	+		−
Eimeria spinifera	−	+	±	+		−
Eimeria vesicostieda	−	−	+		+	−
Eimeria innominata	−	−	−		?	−
Eimeria irregularis	−	−	−	+		−
Eimeria koormae	+	−	−		+	−
Eimeria légeri	−	−	−	+		−
Eimeria triangularis	−	−	−	+		3?
Eimeria trionyxae	−	?	−	+		−
Eimeria lutotestudinis	−	+	−	+		−
Eimeria caretta	−	−	−	+		−
Caryospora cheloniae	−	−	−	+		−
Isospora chelydrae	−	−	−	+		3
Isospora rodriguesae	−	−	+		+	−
Isospora testudae	−	−	−	?		−

(*Continued*)

Table 4 (Continued)

Turtle Suborder/Coccidia Species	Oocyst Structures[a]			Wall Thickness		Wall with Tubercles/ Points
	M	OR	PG	≤1.0	≥1.1	
Pleurodira						
Eimeria jirkamoraveci	−	−	−	+		3
Eimeria lokuma	−	+	−	+		−
Eimeria peltocephali	−	+	−	+		−
Eimeria lagunculata	+	−	−	+		−
Eimeria mammiformis	+	−	±	+		−
Eimeria podocnemis	−	−	+	+		−

[a]M, micropyle; OR, oocyst residuum; PG, polar granule(s); +, present; −, absent.

LITERATURE CITED

Anonymous. 2010. Biology Base.com website. http://biologybase.wikispaces.com/Testudines.

Bailey, R.M. 1941. The occurrence of the wood turtle in Iowa. Copia 1941:265.

Barnard, S.M., Upton, S.J. 1994. *A Veterinary Guide to the Parasites of Reptiles*. Krieger Publishing Co., Malabar, FL. 154 p.

Bhatia, B.L. 1938. *Fauna of British India (Protozoa: Sporozoa)*. Taylor and Francis, Ltd., London, UK. 497 p.

Bone, L.W. 1975. *Eimeria pseudemydis* Lainson, 1968, from the red-eared turtle, *Pseudemys scripta elegans* in Arkansas. Journal of Wildlife Diseases 11:290–291.

Bour, R. 1987. Type-specimen of the alligator snapper, *Macroclemys temminckii* (Harlan, 1935). Journal of Herpetology 21:340–343.

Bourdeau, P. 1988. Diseases of turtles: disease of the skin and digestive tract. Le Point Vétérinaire 20:871–884.

Bourdeau, P. 1989. Pathologie des tortues 2e parte: affections cutanées et digestives (Diseases of turtles Part 2: diseases of the skin and digestive tract). Le Point Vétérinaire 21:19–32 (in French with English abstract).

Carini, A. 1942. Sobre uma *Eimeria* da "*Testudo tabulata*." Arquivos de Biologia, Sao Paulo 26:163–164.

Cerruti, C. 1930. Su di un coccidio parassito di *Testudo graeca*, Linn. Archivio Italiano di Scienzee Medicina Coloiale e di Parassitologia 11:328–331.

Chakravarty, M., Kar, A.B. 1943. Observations on two coccidia, *Eimeria trionyxae* n. sp. and *Eimeria triangularis* n. sp., from the intestine of the turtle *Trionyx gangeticus* Cuv. Journal of the Royal Asiatic Society of Bengal, Science 9:49–54.

Couch, L., Stone, P.A., Duszynski, D.W., Snell, H.L., Snell, H.M. 1996. A survey of the coccidian parasites of reptiles from islands of the Galapagos archipelago: 1990–1994. Journal of Parasitology 82:432–437.

Cranfield, M.R., Graczyk, T.K. 2000. Cryptosporidia in Reptiles. In: *Kirk's Current Veterinary Therapy XIII Small Animal Practice*, Bonagura, J.D. (ed.). W.B. Saunders Company, A Division of Harcourt Brace & Company, Philadelphia, London, Toronto, Montreal, Sydney, Tokyo, Pp. 1188–1191.

Das-Gupta, M. 1938a. On a new coccidium, *Eimeria koormai* n. sp. from an Indian tortoise, *Lissemys punctata* Smith. Proceedings of the Indian Science Congress, Section of Zoology, Abstract No. 2, p. 155.

Das-Gupta, M. 1938b. On a new coccidium *Eimeria koormae* n. sp. from the intestine of Indian tortoise, *Lissemys punctata* Smith. Archiv für Protestenkunde 90:410–413.

Davronov, O. 1985. Coccidia of reptiles from southern Uzbekistan. Parazitologiya (Leningrad) 19:158–161.

Deeds, O.J., Jahn, T.L. 1939. Coccidian infections of western painted turtles of the Okoboji region. Transactions of the American Microscopical Society 58:249–253.

Doflein, F. 1909. *Lehrbuch der Protozoenkunde. Eine Dartstellung der Naturgeschichte der Protozoen mit Besononderer Berucksichtigung der Parasitischen und Pathogenen Formen.* Verlag von Gustav Fischer, Jena. 914 p.

Douglas, R.J., Sundermann, C.A., Lindsay, D.S. 1991. Effects of route of inoculation on the site of development of *Caryospora bigenetica.* Journal of Parasitology 77:755–757.

Douglas, R.J., Sundermann, C.A., Lindsay, D.S., Mulvaney, D.R. 1992. Experimental *Caryospora bigenetica* (Apicomplexa: Eimeriidae) infections in swine (*Sus scrofa*). Journal of Parasitology 78:148–151.

Dubey, J.P., Black, S.S., Sangster, L.T., Lindsay, D.S., Sundermann, C.A., Topper, M.J. 1990. *Caryospora*-associated dermatitis in dogs. Journal of Parasitology 76:552–556.

Duszynski, D.W. 1999. Revisiting the code: clarifying name-bearing types for photomicrographs of Protozoa. Critical comment. Journal of Parasitology 85:769–770.

Duszynski, D.W. 2002. *Coccidia (Apicomplexa: Eimeriidae) of the Mammalian Order Chiroptera.* Special Publication of the Museum of Southwestern Biology, No. 5. University of New Mexico, Albuquerque, NM. 45 p.

Duszynski, D.W., Couch, L. 2013. *The biology and Identification of the Coccidia (Apicomplexa) of Rabbits of the World.* Academic Press (an imprint of Elsevier), Amsterdam, The Netherlands. 340 p.

Duszynski, D.W., Upton, S.J. 2000. *Coccidia (Apicomplexa: Eimeriidae) of the Mammalian Order Insectivora.* Special Publication of the Museum of Southwestern Biology, No 4. University of New Mexico, Albuquerque, NM. 67 p.

Duszynski, D.W., Upton, S.J. 2001. *Cyclospora, Eimeria, Isospora,* and *Cryptosporidium* spp. In: Samuel, W.M., Pybus, J.J., Kocan, A.A. (eds.), *Parasitic Diseases of Wild Mammals,* 2nd ed. Iowa State University Press, Ames, IA, pp. 416–459.

Duszynski, D.W., Upton, S.J. 2009. *The Biology of the Coccidia (Apicomplexa) of Snakes of the World. A Scholarly Handbook for Identification and Treatment.* https://www.CreateSpace.com/3388533. 422 p. ISBN 1448617995.

Duszynski, D.W., Wilber, P.G. 1997. A guideline for the preparation of species descriptions in the Eimeriidae. Journal of Parasitology 83:333–336.

Duszynski, D.W., Upton, S.J., Bolek, M. 2007. Coccidia (Apicomplexa: Eimeriidae) of the amphibians of the world. Zootaxa (Magnolia Press) 1667:1–77.

Duszynski, D.W., Wilson, W.D., Upton, S.J., Levine, N.D. 1999. Coccidia (Apicomplexa: Eimeriidae) in the Primates and the Scandentia. International Journal of Primatology 20:761–797.

Ernst, C.H., Ernst, E.M. 1979. Synopsis of protozoans parasitic in native turtles of the United States. Bulletin of the Maryland Herpetological Society 15:1–15.

Ernst, J.V., Forrester, D.J. 1973. *Eimeria carri* sp. n. (Protozoa: Eimeriidae) from the box turtle, *Terrapene carolina.* Journal of Parasitology 59:635–636.

Ernst, J.V., Fincher, G.T., Stewart, T.B. 1971. *Eimeria paynei* sp. n. (Protozoa: Eimeriidae) from the Gopher tortoise, *Gopherus polyphemus.* Proceedings of the Helminthological Society of Washington 38:223–224.

Ernst, J.V., Stewart, T.B., Sampson, J.R., Fincher, G.T. 1969. *Eimeria chelydrae* n. sp. (Protozoa: Eimeriidae) from the snapping turtle, *Chelydra serpentina.* Bulletin of the Wildlife Disease Association 5:410–411.

Escalante, A.A., Ayala, F.J. 1995. Evolutionary origin of *Plasmodium* and other Apicomplexa based on rRNA genes. Proceedings of the National Academy of Sciences, USA 92:5793–5797.

Frey, J.K., Yates, T.L., Duszynski, D.W., Gannon, W.L., Gardner, S.L. 1992. Designation and curatorial management of type host specimens (symbiotypes) for new parasite species. Journal of Parasitology 78:930–932.

Funk, R.S. 1987. Implications of cryptosporidiosis in Emerald tree boas (*Corallus caninus*). In: Rosenbert, M.J. (ed.), *11th International Herpetological Symposium on Captive Propagation and Husbandry*, Zoological Consortium, Thurmont, MD, Pp. 139–143.

Gaffney, E.S., Hutchison, J.H., Jenkins, F.A., Meeker, L.J. 1987. Modern turtle origins: The oldest known cryptodire. Science 237:289–291.

Garner, M.M., Gardiner, C., Linn, M., McNamara, T.S., Raphael, B., Lung, N.P., Kleinpeter, D., Norton, T.M., Jacobson, E. 1998. Seven new cases of intranuclear coccidiosis in tortoises: An emerging disease? In: Scientific Proceedings, Joint Conference, American Association of Zoo Veterinarians & American Association of Wildlife Veterinarians, Omaha, NE, Pp. 71–73 (AAZV).

Garner, M.M., Gardiner, C.H., Wellehan, J.F.X., Johnson, A.J., McNamara, T., Linn, M., Terrell, S.P., Childress, A., Jacobson, E.R. 2006. Intranuclear coccidiosis in tortoises: nine cases. Veterinary Pathology 43:311–320.

Georges, A., Birrell, J., Saint, K.M., McCord, W., Donnellan, S.C. 1998. A phylogeny for side-necked turtles (Chelonia: Pleurodira) based on mitochondrial and nuclear gene sequence variation. Biological Journal of the Linnean Society 67:213–246.

Gordon, A.N., Kelly, W.R., Lester, J.G. 1993a. Coccidiosis: a fatal disease of free-living green turtles, *Chelonia mydas*. Marine Turtle Newsletter 61:2–3.

Gordon, A.N., Kelly, W.R., Lester, J.G. 1993b. Epizootic mortality of free-living green turtles, *Chelonia mydas*, due to coccidiosis. Journal of Wildlife Diseases 29:490–494.

Graczyk, T.K., Cranfield, M.R. 1998. Experimental transmission of *Cryptosporidium* oocyst isolates from mammals, birds, and reptiles to captive snakes. Veterinary Research 29:187–195.

Graczyk, T.K., Cranfield, M.R., Fayer, R. 1996. Evaluation of commercial enzyme immunoassay (EIA) and immunofluorescent antibody (IFA) tests kits for detection of *Cryptosporidium* oocysts other than *Cryptosporidium parvum*. American Journal of Tropical Medicine and Hygiene 53:274–279.

Graczyk, T.K., Balazs, G.H., Work, T., Aguirre, A.A., Ellis, D.M., Murakawa, S.K.K., Morris, R. 1997. *Cryptosporidium* sp. infections in green turtles, *Chelonia mydas*, as a potential source of marine waterborne oocysts in the Hawaiian Islands. Applied and Environmental Microbiology 63:2925–2927.

Graczyk, T.K., Cranfield, M.R., Hill, S.L. 1996. Therapeutic efficacy of halofuginone and spiramycin treatment against *Cryptosporidium serpentes* (Apicomplexa: Cryptosporidiidae) infections in captive snakes. Parasitology Research 82:143–148.

Graczyk, T.K., Cranfield, M.R., Helmer, P., Fayer, R., Bostwick, E.F. 1998. Therapeutic efficacy of hyperimmune bovine colostrum treatment against clinical and subclinical *Cryptosporidium serpentis* infection in captive snakes. Veterinary Parasitology 74:123–132.

Graczyk, T.K., Cranfield, M.R., Mann, J., Strandberg, J.D. 1998. Intestinal *Cryptosporidium* sp. infection in the Egyptian tortoise, *Testudo kleinmanni*. International Journal for Parasitology 28:1885–1888.

Greiner, E.C. 2003. Coccidiosis in reptiles. Seminars in Avian and Exotic Pet Medicine 12:49–56.

Häfeli, W., Zwart, P. 2000. Panzerweiche bei jungen Landschildkröten und deren mögliche Ursachen. Der Praktische Tierarzt 81:129–132.

Heuschele, W.P., Osterhuis, J., Janssen, D., Robinson, P.T., Ensley, P.K., Meier, J.E., Olson, T., Anderson, M.P., Benirschke, K. 1986. Cryptosporidial infections in captive wild animals. Journal of Wildlife Diseases 22:493–496.

Hůrková, L., Modrý, D., Koudela, B., Šlapeta, J. 2000. Description of *Eimeria motelo* sp. n. (Apicomplexa: Eimeriidae) from the yellow footed tortoise, *Geochelone denticulate* (Chelonia: Testudinidae), and replacement of *E. carinii* Lainson, Costa and Shaw, 1990 by *Eimeria lainsoni* nom. nov. Memorias do Instituto Oswaldo Cruz 95:829–832.

Innis, C.J., Garner, M., Tabaka, C., Greiner, E., Rarschang, R., Gordon, D., Wendland, L. Undated report. Clinical and histopathology findings in Sulawesi tortoises (*Indotestudo forstenii*) with necrotizing sinusitis and rhinitis. Unpublished.

Innis, C.G., Garner, M.M., Johnson, A.J., Wellehan, J.F.X., Tabaka, C., Marschang, R.E., Nordhausen, R.W., Jacobson, E.R. 2007. Antemortem diagnosis and characterization of nasal intranuclear coccidiosis in Sulawesi tortoises (*Indotestudo forstenii*). Journal of Veterinary Diagnostic Investigation 19:660–667.

IUCN, International Union for the Conservation of Nature and Natural Resources. 2009. The IUCN Red List of Threatened Species. Version 2009.1, http://www.iucnredlist.org.

Jacobson, E.R., Schumacher, J., Telford, S.R., Greiner, E.C., Buergelt, C.D., Gardiner, C.H. 1994. Intranuclear coccidiosis in radiated tortoises (*Geochelone radiata*). Journal of Zoo Wildlife Medicine 25:95–102.

Kar, A.B. 1944. Two new coccidia from pond turtles, *Lissemys punctana*—(Bonnaterre). Indian Veterinary Journal 20:231–234.

Keymer, I.F. 1978a. Diseases of chelonians: (1) necropsy survey of tortoises. Veterinary Record 103:548–552.

Keymer, I.F. 1978b. Diseases of chelonians: (2) necropsy survey of terrapins and turtles. Veterinary Record 103:577–582.

Kinne. O. (ed.). 1985. Diseases caused by protistans. In: *Diseases of Marine Animals*. Vol. IV, Part 2, Introduction: Reptilia, Aves, Mammalia. Biologische Anstalt Helgoland, Hamburg, Germany. Pp. 566–571.

Labbé, A. 1893. *Coccidium delagei* coccidie nouvelle parasite des tortues d'eau douce. Archives de Zoologie Experimentale et Generale 1:267–280.

Lainson, R. 1968. Parasitological studies in British Honduras. IV. Some coccidial parasites of reptiles. Annals of Tropical Medicine and Parasitology 62:260–266.

Lainson, R., Naiff, R.D. 1998. *Eimeria peltocephali* n. sp. (Apicomplexa: Eimeriidae) from the freshwater turtle *Peltocephalus dumerilianus* (Chelonia: Pelomusidae) and *Eimeria molossi* n. sp., from the bat, *Molossus ater* (Mammalia: Chiroptera). Memorias do Instituto Oswaldo Cruz 93:81–90.

Lainson, R., Shaw, J.J. 1971. *Sarcocystis gracilis* n. sp. from the Brazilian tortoise *Kinosternon scorpioides*. Journal of Protozoology 18:365–372.

Lainson, R., Shaw, J.J. 1972. *Sarcocystis* in tortoises: a replacement name, *Sarcocystis kinosterni*, for the homonym *Sarcocystis gracilis* Lainson and Shaw, 1971. Journal of Protozoology 19:212.

Lainson, R., Costa, A.M., Shaw, J.J. 1990. *Eimeria* species (Apicomplexa: Eimeriidae) of *Podocnemis expansa* (Schweigger) and *Geochelone denticulata* (Linn.) from Amazonia Brazil (Reptilia: Chelonia). Memorias do Instituto Oswaldo Cruz 85:383–390.

Lainson, R., da Silva, F.M.M., Franco, C.M., de Souza, M.C. 2008. New species of *Eimeria* and *Isospora* (Protozoa: Eimeriidae) in *Geochelone* spp. (Chelonia: Testudinidae) from Amazonia Brazil. Parasite 15:531–538.

Laveran, A., Mesnil, F. 1902. Sur quelques protozoaires parasites d'une tortue d'Asie (*Damonia reevesii*). Comptes Rendus de l'Academie des Sciences, Series 3 134, 609–614.

Leibovitz, L., Rebell, G., Boucher, G.C. 1978. *Caryospora cheloniae* sp. n.: a coccidial pathogen of mariculture-reared green sea turtles (*Chelonia mydas mydas*). Journal of Wildlife Diseases 14:269–275.

Lindsay, D.S., Sundermann, C.A. 1989. Recent advances in the biology of the coccidian genus *Caryospora*. Journal of Veterinary Parasitology 3:1–5.

Lauckner, G. 1985. Chapter 2. Diseases of reptiles. In: Kinne, O. (ed.), *Diseases of Marine Animals*, Vol. IV, Part 2, Biologische Anstalt Helgoland, Hamburg, Germany, Pp. 552–626.

Mandal, A.K. 1976. Coccidia of Indian vertebrates. Records of the Zoological Survey of India 70:39–120.

McAllister, C.T. 1989. Systematics of Coccidian Parasites (Apicomplexa) from Amphibians and Reptiles in Northcentral Texas. Ph.D. Dissertation, University of North Texas, Denton, TX. 152 p.

McAllister, C.T., Upton, S.J. 1988. *Eimeria trachemydis* n. sp. (Apicomplexa: Eimeriidae) and other eimerians from the red-eared slider, *Trachemys scripta elegans* (Reptilia: Testudines), in North-Central Texas. Journal of Parasitology 74:1014–1017.

McAllister, C.T., Upton, S.J. 1989a. *Eimeria ornata* n. sp. (Apicomplexa: Eimeriidae) from the ornate box turtle, *Terrapene ornata ornata* (Reptilia: Testudines), in Texas. Journal of Protozoology 36:131–133.

McAllister, C.T., Upton, S.J. 1989b. The coccidia (Apicomplexa: eimeriidae) of Testudines, with descriptions of three new species. Canadian Journal of Zoology 67:2459–2467.

McAllister, C.T., Upton, S.J. 1992. A new species of *Eimeria* (Apicomplexa: Eimeriidae) from *Pseudemys texana* (Testudines: Emydidae), from North-Central Texas. Texas Journal of Science 44:37–41.

McAllister, C.T., Duszynski, D.W., Roberts, D.T. 2014. A new coccidian (Apicomplexa: Eimeriidae) from Galápagos tortoise, *Chelonoidis* sp. (Testudines: Testudinidae), from the Dallas Zoo. Journal of Parasitology 100:128–132.

McAllister, C.T., Stuart, J.N., Upton, S.J. 1995. Coccidia (Apicomplexa: Eimeriidae) from the Big Bend Slider, *Trachemys gaigeae* (Testudines: Emydidae), in New Mexico. Journal of Parasitology 81:804–805.

McAllister, C.T., Upton, S.J., McCaskill, L.D. 1990a. Three new species of *Eimeria* (Apicomplexa: Eimeriidae) from *Apalone spinifera pallidus* (Testudines: Trionychidae) in Texas, with a redescription of *E. amydae*. Journal of Parasitology 76:481–486.

McAllister, C.T., Upton, S.J., Trauth, S.E. 1990b. Coccidian parasites (Apicomplexa: Eimeriidae) of *Chelydra serpentina* (Testudines: Chelydridae) from Arkansas and Texas, U.S.A., with descriptions of *Isospora chelydrae* sp. nov. and *Eimeria serpentina* sp. nov. Canadian Journal of Zoology 68:865–868.

McAllister, C.T., Upton, S.J., Trauth, S.E. 1990c. A new species of *Eimeria* (Apicomplexa: Eimeriidae) from the green water snake, *Nerodia cyclopion* (Reptilia: Serpentes), in Arkansas, U.S.A. Transactions of the American Microscopical Society 109:69–73.

McAllister, C.T., Upton, S.J., Trauth, S.E. 1994. New host and geographic records for coccidia (Apicomplexa: Eimeriidae) from North American turtles. Journal of Parasitology 80:1045–1049.

McAllister, C.T., Upton, S.J., Killebrew, F.C. 1991. Coccidian parasites (Apicomplexa: Eimeriidae) of *Graptemys cageli* and *G. versa* (Testudines: Emydidae) from Texas. Journal of Parasitology 77:500–502.

Meshkov, S. 1975. Sarcosporidia among tortoises in south-eastern Bulgaria. Doklady Bulgarska Akademiia Na Naukite and/or Comptes rendus de l'Académie Bulgare des Sciences 28:1547–1548.

Novilla, M.N., Carpenter, J.W., Spraker, T.R., Jeffers, T.K. 1981. Parental (*sic*) development of eimerian coccidia in sandhill and whooping cranes. Journal of Parasitology 28:248–255.

Overstreet, R.M. 1981. Species of *Eimeria* in nonepithelial sites. Journal of Parasitology 28:258–260.

Ovezmukhammedov, A. 1978. Coccidiofauna of *Emys orbicularis* Linnaeus in Turkmenistan. Izvestiia Akademii Nauk Turkmenia SSR seriya Biologischeskikh Nauk 0 (No. 1):83–86 (in Russian).

Paperna, I., Landsberg, J.H. 1989. Description and taxonomic discussion of eimerian coccidia from African and Levantine geckoes. South African Journal of Zoology 24:345–355.

Pavlásek, I. 1998. Některé nové poznatky o kryptosporidiích u plazů chovaných v zajetí v České republice (Some new knowledge about Cryptosporidia in captive reptiles in the Czech Republic). Gazella 25:163–170 (in Czech, English summary).

Pellérdy, L.P. 1963. *Catalogue of Eimeriidea (Protozoa: Sporozoa)*. Publishing House of the Hungarian Academy of Science, Akademia Kiado, Budapest, Hungary. 160 p.

Pellérdy, L.P. 1974. *Coccidia and Coccidiosis*, 2nd ed. Verlag Paul Parey, Berlin and Hamburg, and Akademia Kiado, Budapest, Hungary. 959 p.

Pinto, C. 1928. *Eimeria carinii* nova especie. Parasita de *Mus* (E.) *norvegicus* do Brazil. Bol. Biol. Sao Paulo 11–14:127–128.

Pluto, T.G., Rothenbacher, H. 1976. *Eimeria juniataensis* sp. n. (Protozoa: Eimeriidae) from the map turtle, *Graptemys geographica*, in Pennsylvania. Journal of Parasitology 62:207–208.

Pough, F.H., Andrews, R.M., Cadle, J.E., Crump, M.L., Savitsky, A.H., Wells, K.D. 2004. *Systematics and diversity of extant reptiles, Ch. 4, Herpetology*, 3rd ed. Pearson, Prentice Hall, Upper Saddle River, NJ, pp. 97–173.

Poulin, R., Morand, S. 2000. The diversity of parasites. Quarterly Review of Biology. 75:277–293.

Rebell, G., Rywlin, A., Ulrich, G. 1974. Coccidiosis in the green turtle (*Chelonia mydas*) in mariculture. In: Avault, Jr., J.W. (ed.), Proceedings of the 5th Annual Meeting of the World Mariculture Association, Charleston, SC, January 21–25, 1974. Division of Continuing Education, Louisiana State University, Baton Rouge, LA. Pp. 197–204.

Reichenow, E. 1921. Die Coccidia. In: von Prowazek, S.J.M. (ed.), *Hanbuch der Pathogenen Protozoen*, Johann Ambrosius Barth, Leipzig, Germany, Pp. 1136–1277.

Rhodin, A.G.J., van Dijk, P.P., Iverson, J.B., Shaffer, H.B. [Turtle Taxonomy Working Group]. 2010. Turtles of the world, 2010 update: Annotated checklist of taxonomy, synonymy, distribution, and conservation status. In, Conservation Biology of Freshwater Turtles and Tortoises: A Compilation Project of the IUCN/SSC Tortoise and Fresh Water Turtle Specialist Group.

Rhodin, A.G.J., Pritchard, P.C.H., van Dijk, P.P., Saumure, R.A., Buhlmann, K.A., Iverson, J. B., et al. eds. Chelonian Research Monograph, No. 5. 000.85–000.164. http://www.iucn-tftsg.org/cbftt.

Roudabush, R.L. 1937. Some coccidia of reptiles found in North America. Journal of Parasitology 23:354–359.

Sampson, J.R., Ernst, J.V. 1969. *Eimeria scriptae* n. sp. (Sporozoa: Eimeriidae) from the red-eared turtle *Pseudemys scripta elegans*. Journal of Protozoology 16:444–445.

Segade, P., Crespo, C., Ayres, C., Cordero, A., Arias, M.C., Garcia-Estévez, M., Iglesias-Blanco, R. 2006. *Eimeria* species from the European pond turtle, *Emys orbicularis* (Reptilia: Testudines), in Galicia (NW Spain), with description of two new species. Journal of Parasitology 92:69–72.

Simond, P.-L. 1901. Note sur une Coccidia nouvelle, *Coccidium legeri*, Parasite de *Cryptopus granosus* (*Emyda granosa*). Comptes Rendus des Sciences de la Societe de Biologie 53:485–486.

Široký, P., Modrý, D. 2005. Two new species of *Eimeria* (Apicomplexa: Eimeriidae) from Asian geoemydid turtles *Kachuga tentoria* and *Melanochelys trijuga* (Testudines: Geoemydidae). Parasite 12:9–13.

Široký, P., Modrý, D. 2006. Two eimerian coccidia (Apicomplexa: Eimeriidae) from the critically endangered Arakan Forest turtle *Heosemys depressa* (Testudines: Geoemydidae), with description of *Eimeria arakanensis* n. sp. Acta Protozoologica 45:183–189.

Široký, P., Modrý, D. 2010. Eimeriid coccidia (Apicomplexa: Eimeriidae) from geoemydid turtles (Testudines: Geoemydidae) with a description of six new species. Acta Protozoologica 49:301–310.

Široký, P., Kamler, M., Modrý, D. 2006a. A new *Eimeria* (Apicomplexa: Eimeriidae), possessing mitra-shaped oocyst, from the Neotropical chelid turtle *Batrachemys heliostemma* (Testudines: Chelidae), and its comparison with *Eimeria mitraria* (Laveran & Mesnil 1902). Memorias do Instituto Oswaldo Cruz 101:555–558.

Široký, P., Kamler, M., Modrý, D. 2006b. *Eimeria lokuma* n. sp. (Apicomplexa: Eimeriidae), a new coccidium from the African helmeted turtle *Pelomedusa subrufa* (Lacépède) (Testudines: Pelomedusidae). Systematic Parasitology 65:73–76.

Spawls, S., Howell, K., Drewes, R., Ashe, J. 2002. *A Field Guide to the Reptiles of East Africa*. Academic Press, Hong Kong. 543 p.

Stankovitch, S. 1920. Sur deux nouvelles coccidies parasites des poisons cyprinides. Comptes Rendus des Sciences de la Societe de Biologie 83:833–835.

Stone, W.B., Manwell, R.D. 1969. Toxoplasmosis in cold-blooded hosts. Journal of Protozoology 16:99–102.

Sunderman, C.J. 1988. Coccidian disease of green turtles. In: Sundermann, C.J., Lightner, C.V. (eds.), *Disease Diagnosis and Control in North American Marine Aquaculture*, Elsevier Science Publishers, Amsterdam, The Netherlands, Pp. 384–385.

Upton, S.J., Odell, D.K., Walsh, M.T. 1990. *Eimeria caretta* sp. nov. (Apicomplexa: Eimeriidae) from the loggerhead sea turtle, *Caretta caretta* (Testudines). Canadian Journal of Zoology 68:1268–1269.

Upton, S.J., McAllister, C.T., Garrett, C.M. 1995. New species of *Eimeria* (Apicomplexa) from captive wood turtles, *Clemmys insculpta* (Testudines: Emydidae), from the Dallas Zoo. Acta Protozoologica 34:57–60.

Upton, S.J., McAllister, C.T., Trauth, S.E. 1992. Description of a new species of *Eimeria* (Apicomplexa: Eimeriidae) from the alligator snapping turtle, *Macroclemys temminckii* (Testudines: Chelydridae). Journal of the Helminthological Society of Washington 59:167–169.

Uetz, P., Hošek, J. (eds.). 2013. *The Reptile Database*. World Wide Web Publication. http://www .reptiledatabase.org/. Accessed daily, April, 2012 through April, 2014.

Wacha, R.S., Christiansen, J.L. 1974a. *Eimeria megalostiedai* sp. n. (Protozoa: Sporozoa) from the wood turtle, *Clemmys insculpta* in Iowa. Proceedings of the Helminthological Society of Washington 41:35–37.

Wacha, R.S., Christiansen, J.L. 1974b. Some coccidian parasites from Iowa turtles. In: Moner, J.G. (ed.), *Proceedings of the 49th Annual Meeting of the American Society of Parasitologists*, Kansas City, Kansas USA, August 4–9, 1974, Allen Press, Lawrence, Kansas, USA. pp. 45 (Abstract 116).

Wacha, R.S., Christiansen, J.L. 1976. Coccidian parasites from Iowa turtles: systematics and prevalence. Journal of Protozoology 23:57–63.

Wacha, R.S., Christiansen, J.L. 1977. Additional notes on the coccidian parasites of the soft-shell turtle, *Trionyx spiniferus* Le Sueur, in Iowa, with a description of *Eimeria vesicostieda* sp. n. Journal of Protozoology 24:357–359.

Wacha, R.S., Christiansen, J.L. 1979a. *Eimeria filamentifera* sp. n. from the snapping turtle, *Chelydra serpentina* (Linné.), in Iowa. Journal of Protozoology 26:353–354.

Wacha, R.S., Christiansen, J.L. 1979b. Taxonomic and ecological considerations in the classification of the coccidian parasites (genus *Eimeria*) of turtles, with a review of those species from turtles of the United States. Proceedings of the 1978 Reptile Disease Symposium. Special Publication, Wildlife Disease Association, Ames, IA. (Wacha and Christiansen, 1979a, listed this reference as *in press*, but we have not been able to find this reference in any library on on-line resource).

Wacha, R.S., Christiansen, J.L. 1980. Coccidian parasites from Iowa turtles. Journal of Protozoology 27 (Abstract 68):28A–29A.

Weishaar, I., Frank, W., Kern, G., von Nickisch, M. 1988. Landschildkröten Europas und Südamerikas als Zwischenwirte für *Sarcocystis*. Mitteilungen–Österreichischen Gesellschaft für Tropenmedizin und Parasitologie 10:95–102.

Wilber, P.G., Duszynski, D.W., Upton, S.J., Seville, R.S., Corliss, J.O. 1998. A revision of the taxonomy and nomenclature of the *Eimeria* spp. (Apicomplexa: Eimeriidae) from rodents in the Tribe Marmotini (Sciuridae). Systematic Parasitology 39:113–135.

Windsor, D.A. 1998. Most of the species on Earth are parasites. International Journal of Parasitology 28:1939–1941.

Xiao, L., Ryan, U.M., Graczyk, T.K., Limor, J., Li, L., Kombert, M., Junge, R., Sulaiman, I.M., Zhou, L., Arrowood, M.J., Koudela, B., Modrý, D., Lal, A.A. 2004. Genetic diversity of *Cryptosporidium* spp. in captive reptiles. Applied and Environmental Microbiology 70:891–899.

Zhao, X., Duszynski, D.W. 2001a. Molecular phylogenies suggest the oocyst residuum can be used to distinguish two independent lineages of *Eimeria* spp. in rodents. Parasitology Research 87:638–643.

Zhao, X., Duszynski, D.W. 2001b. Phylogenetic relationships among rodent *Eimeria* species determined by plastid ORF470 and nuclear 18S rDNA sequences. International Journal of Parasitology 31:715–719.

Zhao, X., Duszynski, D.W., Loker, E.S. 2001. Phylogenetic position of *Eimeria antrozoi*, a bat coccidium (Apicomplexa: Eimeriidae) and its relationship to morphologically similar *Eimeria* spp. from bats and rodents based on nuclear 18S and plastid 23S rDNA sequences. Journal of Parasitology 87:1120–1123.

Printed and bound by CPI Group (UK) Ltd, Croydon, CR0 4YY

03/10/2024

01040427-0004